THE HISTORY OF STRING THEORY AND LOOP QUANTUM GRAVITY

A JOURNEY THROUGH IDEAS, GEOMETRY, AND REALITY

WHERE STRINGS VIBRATE, GEOMETRY CONNECTS, AND REALITY EMERGES.

FEATURING THE PRIMACOHEDRON: A BRIDGE BETWEEN GEOMETRY AND THE FABRIC OF THE COSMOS

SANDI SETIAWAN

Life Member of Clare Hall, University of Cambridge

PREFACE

A Journey Through Ideas, Geometry, and Reality

1
From Curiosity To Cosmos

We begin with wonder—the ancient questions that first turned our gaze toward the stars.

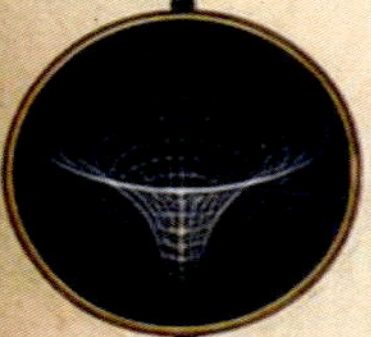

2
The Path of Discovery

Through centuries of insight and imagination, humanity uncovered the deep structure of space, time, and matter.

3
Strings, Loops, and Beyond

String Theory reveals a world of vibrating strings, extra dimensions, and symmetries. Loop Quantum Gravity reveals a world woven from discrete threads of space, where geometry itself is quantum.

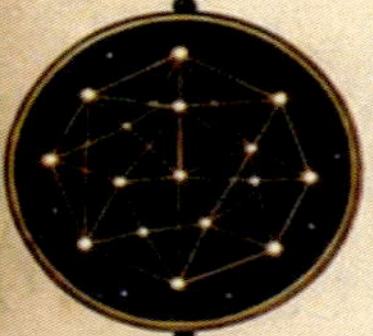

4
The Primacohedron

A bridge between geometry and physics—a hidden harmony that may illuminate the fabric of the universe.

5
Looking Forward

The journey continues. With each step, we move closer to a deeper, more unified understanding of reality.

This book is an invitation—to explore one of the most profound and beautiful quests in modern physics: the quest to understand the fundamental nature of spacetime.

From the smooth continuum of Einstein's relativity to the tiny, vibrating strings of the quantum world, science has journeyed far. Yet, at the intersection of the very large and the very small, we still lack a complete theory.

String Theory and Loop Quantum Gravity are two of the most compelling approaches to this challenge. String Theory envisions elementary particles as vibrational modes of one-dimensional strings, requiring extra dimensions and rich symmetry to reveal the unity of forces and matter. Loop Quantum Gravity, on the other hand, begins with spacetime itself—quantized and discrete—where geometry emerges from networks of loops, woven together in a spin network.

Though different in language and approach, both theories seek to reconcile quantum mechanics and general relativity—to reveal the deep structure of the cosmos.

This journey has been crafted to guide you—step by step—through the history, ideas, and breakthroughs that have shaped this remarkable field. Each page is a doorway: to knowledge, to wonder, and to new questions.

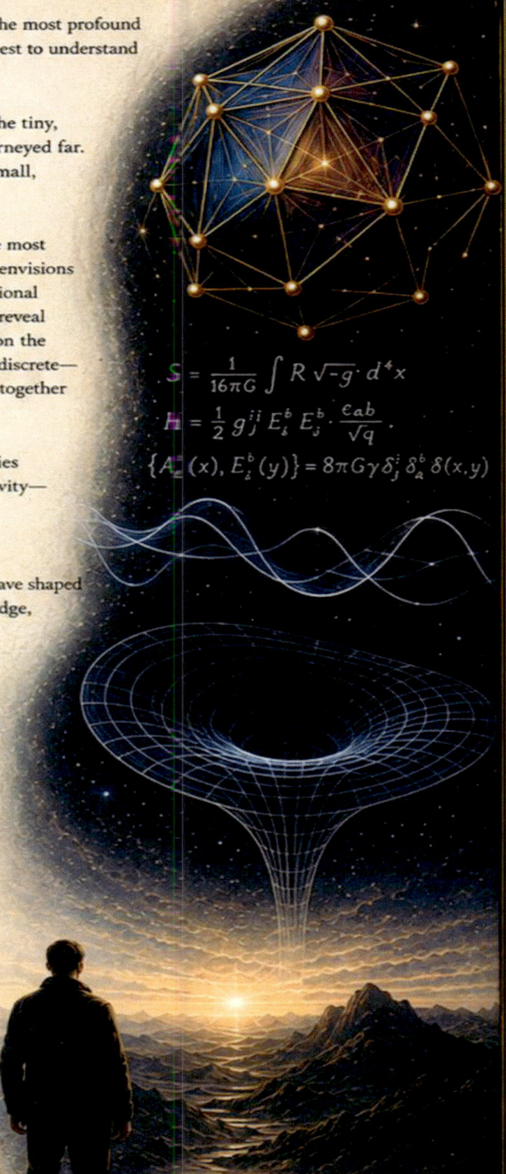

What You Will Find Inside

- A rich history of ideas—from classical physics to the frontiers of String Theory and Loop Quantum Gravity.
- The mathematical and conceptual milestones that led us here.
- A deep dive into String Theory: strings, branes, dualities, and extra dimensions.
- A deep dive into Loop Quantum Gravity: loops, spin networks, quantization of geometry, and the fabric of space.
- The Primacohedron and its role as a geometric bridge between the two paths.
- The challenges that remain, and the exciting future that lies ahead.

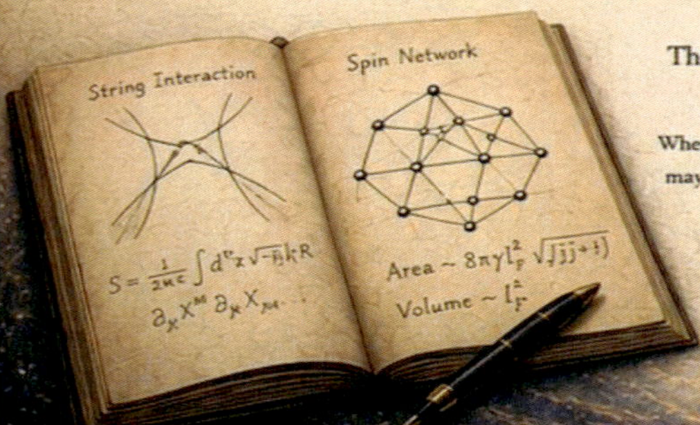

This book is for the curious, the thinkers, the dreamers, and the doers.

Whether you are new to the subject or deep in the research, may it inspire you to keep questioning, keep exploring, and keep imagining.

The universe is still speaking. Let us keep listening.

> *"The important thing is not to stop questioning. Curiosity has its own reason for existing."*
>
> — Albert Einstein

May this journey through the ideas of yesterday light the path for discoveries of tomorrow.

THE HISTORY OF STRING THEORY

1. THE QUEST FOR SEEDS
From Ancient Ideas to Modern Physics

THE BIG QUESTION

For centuries, humans have asked: what is everything made of? The journey to answer this question eventually led to the idea that the fundamental building blocks of nature might not be tiny particles, but tiny vibrating strings.

ANCIENT ROOTS

Greek philosophers such as Thales and Democritus proposed that matter is made of tiny, indivisible units.

19TH & EARLY 20TH CENTURY

Scientific discoveries of atoms, electrons, and quantum phenomena revealed a universe much stranger and richer than anyone had imagined.

MODERN PHYSICS EMERGES

Relativity showed us the fabric of space and time. Quantum mechanics revealed the behavior of the very small. But uniting these two revolutions remained an enormous challenge.

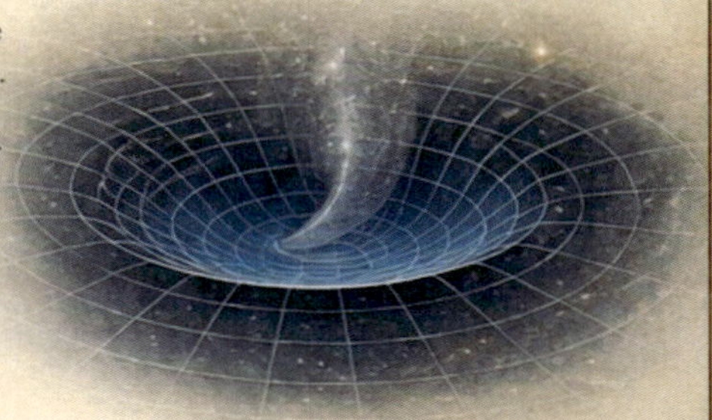

THE SEARCH CONTINUES

From ancient philosophy to modern science, each step brought us closer to a deeper understanding of the universe—and opened the door to a bold new idea: strings as the fundamental constituents of reality.

A TIMELINE OF WONDER

Ancient Ideas
~500 BCE and earlier
Early philosophers wonder about the nature of matter.

19th Century
1800s
Atoms and subatomic particles are discovered.

Early 20th Century
1900s
Relativity and quantum mechanics reshape our understanding.

The Quest Continues
Today and Beyond
The search for a "Theory of Everything" leads us to string theory.

2. Early 20th Century Seeds

Relativity and Quantum Theory Emerge

The early 20th century was a time of revolutionary ideas. Two giants—relativity and quantum theory—began to reshape our understanding of space, time, and the nature of reality.

"The important thing is not to stop questioning. Curiosity has its own reason for existing."

— Albert Einstein

Relativity: A New View of Space and Time

Einstein's theories of relativity revealed that space and time are not absolute, but are woven together into a dynamic fabric that can bend and stretch in response to mass and energy.

Key Idea: Spacetime Curvature

Mass and energy tell spacetime how to curve; curved spacetime tells mass and energy how to move.

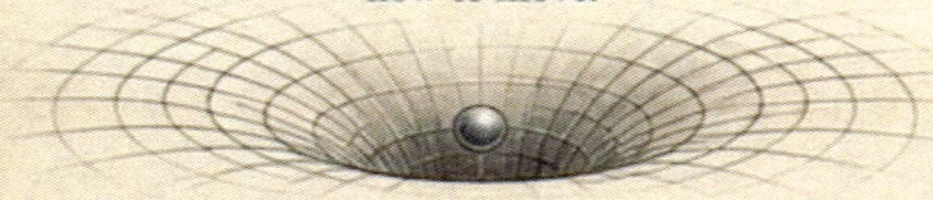

Quantum Theory: The World of the Very Small

Quantum mechanics unveiled a bizarre and fascinating realm where particles can exist in multiple states, be entangled across distances, and behave like both particles and waves.

Key Features

- Quantization of energy
- Wave–particle duality
- Uncertainty principle
- Probabilistic outcomes

"God does not play dice with the universe."

— Albert Einstein

The Growing Tension

Relativity works beautifully for the very large (cosmology, gravity), while quantum theory works for the very small (atoms, particles).

But when we try to unite them, the equations break down. A deeper theory is needed.

?

A Unified Theory?

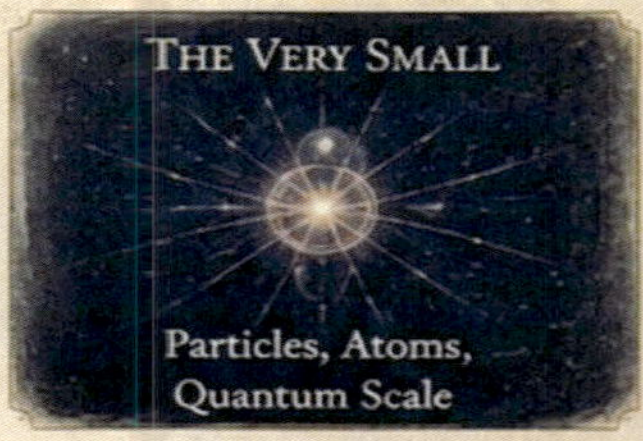

Milestones of the Era

1905
Einstein's Annus Mirabilis
Special relativity, the photoelectric effect, and Brownian motion change physics forever.

1915
General Relativity
Gravity is geometry. Einstein publishes his theory of gravitation.

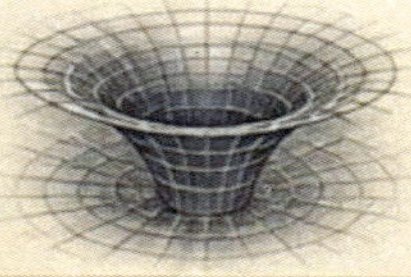

1920s
Quantum Mechanics Takes Shape
Heisenberg, Schrödinger, Dirac, and others build the framework of quantum theory.

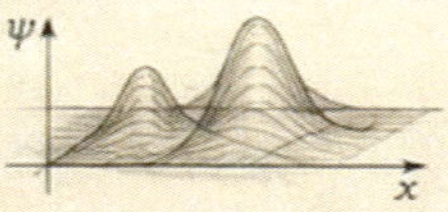

1927
Uncertainty Principle
Heisenberg shows there are limits to what we can know about nature.

$$\Delta x \cdot \Delta p \geq \frac{\hbar}{2}$$

1930s
New Questions Arise
Physicists seek a deeper understanding to unify the two revolutions.

The seeds are planted. The search for unity has begun.

THE HISTORY OF STRING THEORY

3

3. THE S-MATRIX PROGRAM

1960s

In the 1960s, physicists focused on understanding how subatomic particles scatter and interact. The S-matrix program aimed to describe all particle interactions in terms of scatterings, without relying on a deeper theory of what happens at very small scales.

Particles and Forces

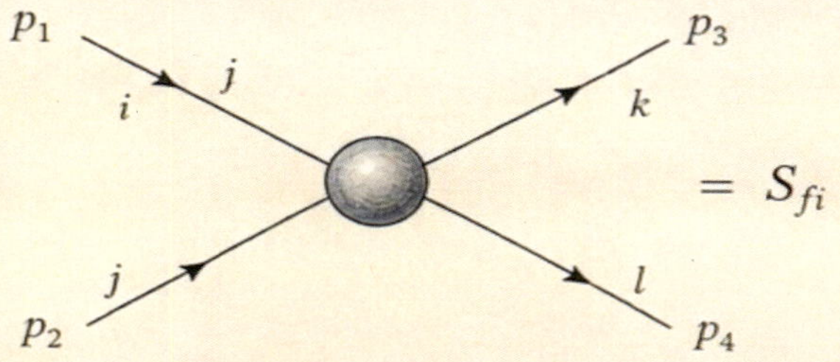

p_i = initial particles
p_f = final particles
S_{fi} = scattering amplitude

Physicists explored scattering amplitudes to understand strong interactions and discover patterns that hinted at a deeper, underlying structure.

THE GOAL

Build a consistent framework where all possible scattering processes could be calculated and related—without needing to know the details of the underlying forces at very short distances.

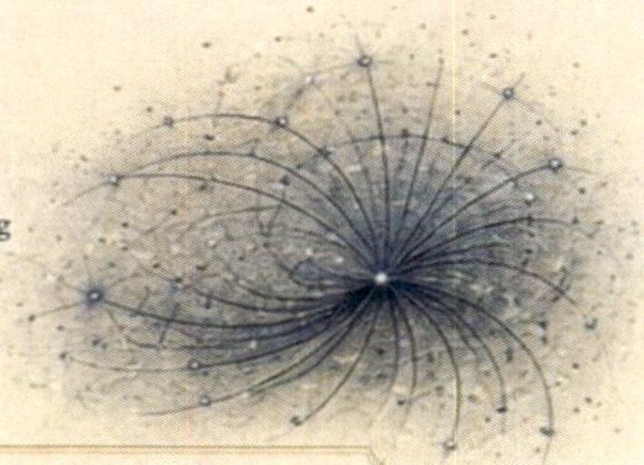

"We are looking for the principles that determine the S-matrix."
— Geoffrey Chew

KEY CONTRIBUTIONS

- ★ **Dispersion Relations**
  Linked scattering amplitudes at different energies using principles of causality and unitarity.
- ★ **Analyticity and Unitarity**
  Showed that amplitudes must be smooth and conserve probability.
- ★ **Bootstrap Ideas**
  Explored the possibility that consistency alone could determine the strong force.

Geoffrey Chew
One of the leaders of the S-matrix program.

WHAT THEY DISCOVERED

Patterns emerged in scattering data that suggested deeper symmetries and hidden structures in nature.

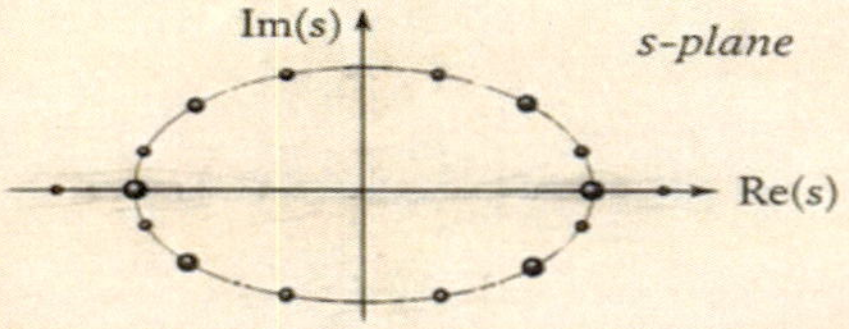

These insights laid the groundwork for future breakthroughs—including string theory.

A STEP TOWARD UNITY

The S-matrix program didn't find a final theory, but it revealed that the universe might be far more interconnected and elegant than expected. It opened the door to the next giant leap: reimagining particles not as points, but as strings.

TIMELINE SNAPSHOT

- **1950s** — S-matrix ideas begin to form
- **1960–1962** — Dispersion relations and analyticity established
- **Mid-1960s** — S-matrix program flourishes
- **Late 1960s** — Limits of the program become clear
- **1970s** — New ideas lead toward string theory

THE HISTORY OF

STRING THEORY

4

4. A NEW IDEA EMERGES

1968

Veneziano Amplitude

In 1968, physicist Gabriele Veneziano proposed a surprising mathematical formula that described how particles scatter at high energies.

It didn't look like anything from quantum field theory—yet it worked beautifully and hinted at a hidden simplicity.

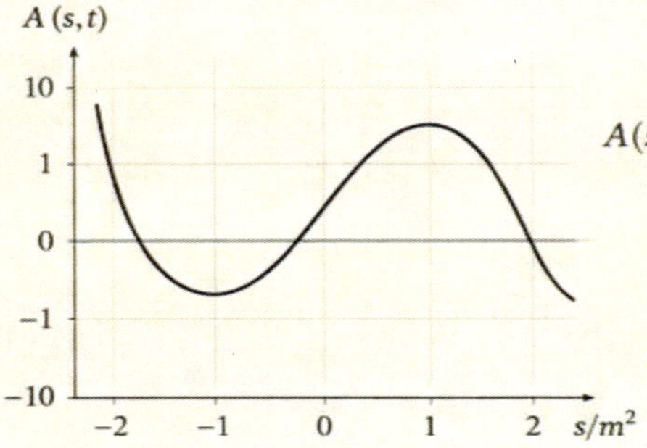

$$A(s,t) = \frac{\Gamma(-\alpha(s))\,\Gamma(-\alpha(t))}{\Gamma(-\alpha(s)-\alpha(t))}$$

$\Gamma(z)$ is the gamma function

$\alpha(s) = \alpha_0 + \alpha' s$ is a linear function called the *Regge trajectory.*

GABRIELE VENEZIANO

His formula was the first clue that strings might underlie the interactions of particles.

WHAT THIS IDEA SUGGESTED

RESONANCES WITHOUT POINTS

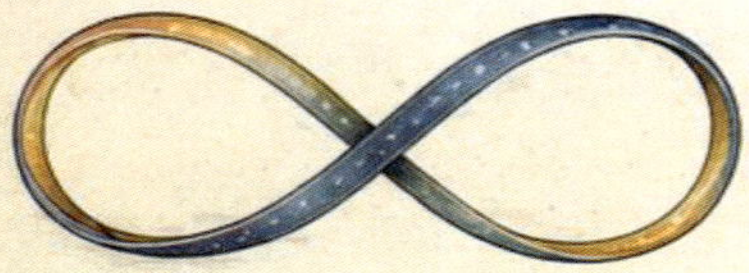

The pattern of resonances matched the vibrations of a string, not the collisions of point particles.

A RADICALLY NEW VIEW

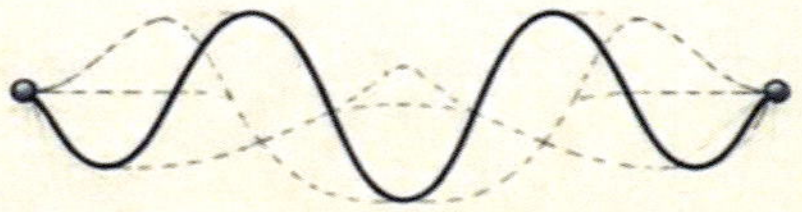

Particles could be different modes of vibration of a single one-dimensional object.

WHY IT MATTERED

- ★ It reproduced known particle data with remarkable accuracy.
- ★ It hinted at an underlying unity in nature.
- ★ It planted the seed of a new theory: strings instead of points.

THE IMPLICATION

If this idea was right, then space, time, and matter might emerge from the vibrations of tiny strings living in a universe with more than the familiar dimensions we perceive.

> *"The most beautiful experience we can have is the mysterious."*
>
> — Albert Einstein

MILESTONES LEADING TO CHANGE

1961
Dual Models
Early models described hadrons as extended objects, not points.

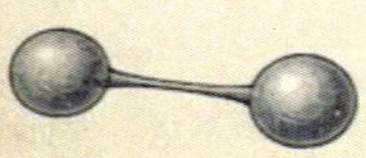

1967
Regge Poles
Ideas from complex analysis revealed deep patterns in scattering.

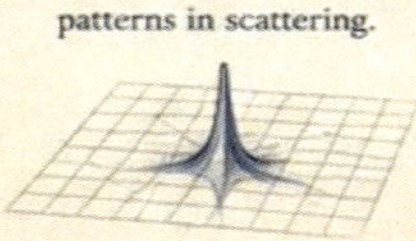

1968
Veneziano Formula
A single equation fit the data perfectly—too well to be just a coincidence.

Late 1960s
String Hypothesis
The formula inspired the idea that strings explain these patterns.

Early 1970s
Toward a Theory
Many physicists began building the foundations of string theory.

☆ *A simple formula opened the door to a profound idea: The universe may be woven from strings, not points.* ☆

THE HISTORY OF STRING THEORY

5

5. STRINGS INSTEAD OF POINTS

LATE 1960s

A Radical Proposal

By the late 1960s, a bold idea took shape: what if the fundamental building blocks of nature are not point particles, but tiny vibrating strings?

This radical proposal offered a new way to understand the particles and forces of the universe.

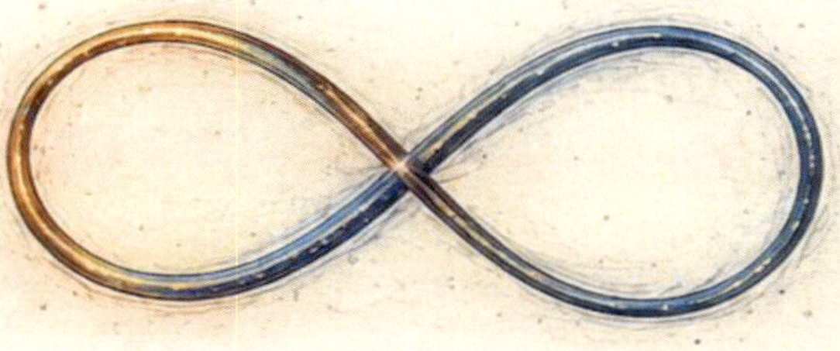

Vibrating strings can produce different particles, depending on how they vibrate.

> *"If we replace points by one-dimensional objects, new possibilities emerge."*
>
> — Leonard Susskind

WHY STRINGS?

1. SOLVES INFINITIES

Strings have a length, not zero size. This smooths out interactions and avoids the infinities that plagued point-particle theories.

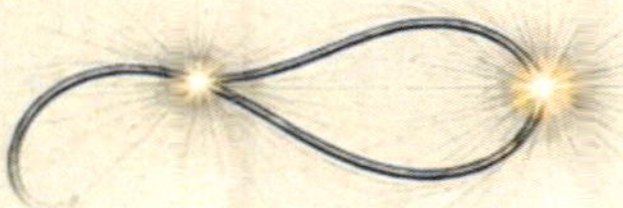

2. UNIFIES PARTICLES & FORCES

Different vibration modes of a string naturally produce different particles—including the forces that act on them.

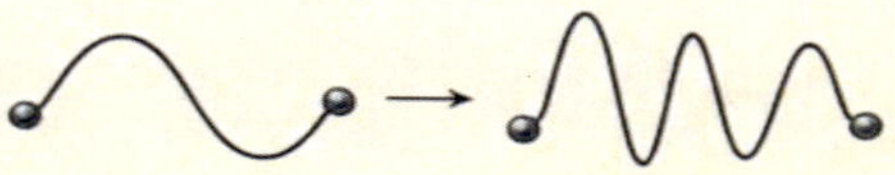

Low vibration (e.g., graviton)

High vibration (e.g., photon)

3. PREDICTS NEW WORLD

Strings hint at extra dimensions and a deeper, more elegant structure of reality.

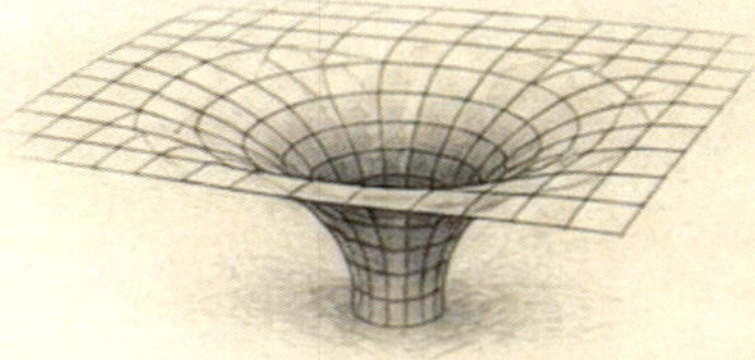

PIONEERS OF THE IDEA

- ★ Leonard Susskind
- ★ Holger Bech Nielsen
- ★ Yoichiro Nambu

and many others

A CHANGE IN PARADIGM

Particles are like notes played by an instrument.

Strings are the instrument.

A new language for nature.

KEY MOMENTS

1958
Early Ideas
The concept of using one-dimensional objects emerges.

1969
Dual Resonance Model
The first model where strings reproduce known particle interactions.

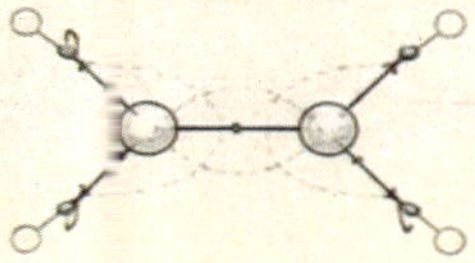

1969–1970
Strings Replace Points
The idea gains momentum—what if all particles are excitations of strings?

Late 1960s
A New Direction
The path toward a unified theory of everything begins.

The Road Ahead
A revolution in physics has begun—one that may reshape our understanding of reality.

> *"The universe may be more like a symphony of strings than a collection of points."*
>
> — Leonard Susskind

The History of String Theory

6

6. Dual Resonance Model

1969

The First String Model

In 1969, Leonard Susskind, Holger Nielsen, and Yoichiro Nambu developed the dual resonance model—an early theory in which particles arise as vibrational modes of one-dimensional strings.

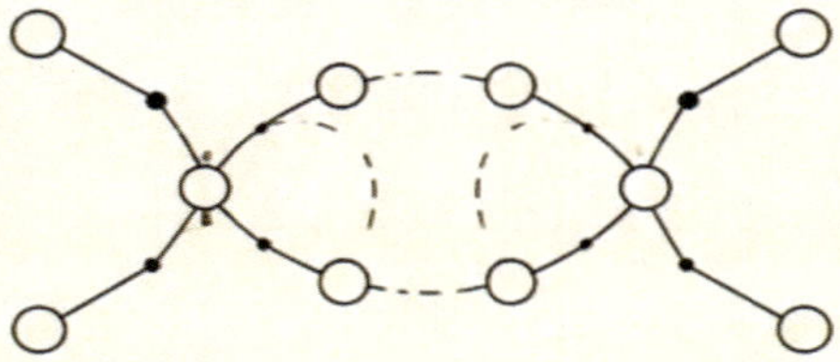

- ○ External particles
- ● Interaction points
- — String (one-dimensional)
- - - - Virtual intermediate states

Particles interact through the exchange of vibrating strings.

The Breakthrough

Leonard Susskind, Holger Nielsen, and Yoichiro Nambu

Their work revealed a surprising duality between seemingly different scattering processes.

Duality: Two Views, One Truth

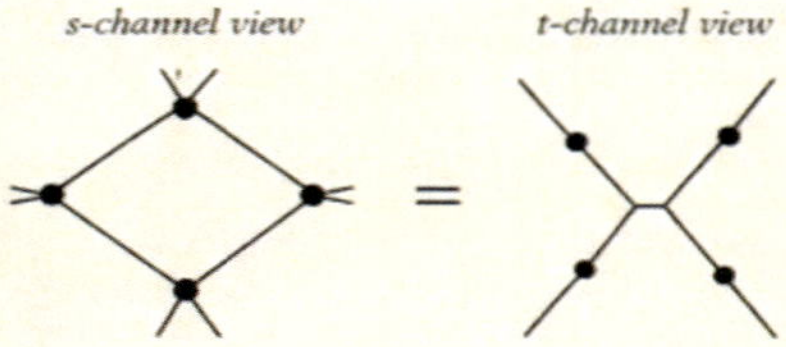

Two different ways of looking at the same process give identical results.

This "duality" was a profound clue that strings, not points, might be fundamental.

What It Achieved

★ Provided the first consistent model where particles are not points but tiny vibrating strings.

★ Naturally incorporated an infinite tower of particles with increasing mass.

★ Offered hope for a unified description of all interactions.

The Beauty of the Model

1. Vibrating String

A string can vibrate in many ways.

Each vibrational mode corresponds to a different particle.

2. Particles From Vibrations

Just as a violin string produces different notes, a fundamental string produces different particles.

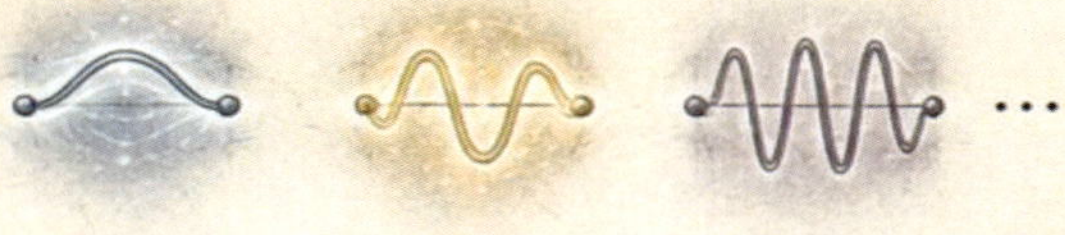

Low mode → lighter particle

Higher mode → heavier particle

Even higher mode → even heavier particle

3. Infinite Tower

The theory predicts an infinite number of possible particles.

Why It Was Revolutionary

- It avoided the infinities plaguing point-particle theories.
- It suggested a deep and hidden symmetry of nature (duality).
- It laid the foundation for everything that followed in string theory.

"The dual resonance model was the spark that ignited the string theory revolution."

The Road Ahead

This first string model was just the beginning. It pointed the way to:

★ higher dimensions

★ supersymmetry

★ richer mathematics

★ and ultimately, a deeper understanding of the universe.

"We were looking for the principles that determine the S-matrix."

— Geoffrey Chew

The History of String Theory

7

7. The Birth of String Theory

Early 1970s

From Dual Models to Strings

In the early 1970s, physicists realized that the dual resonance model could be interpreted in a radical new way: the fundamental building blocks were not point particles, but tiny vibrating one-dimensional objects—strings.

This insight gave birth to string theory, a bold candidate for a unified theory of all fundamental forces.

Vibrating strings can produce the rich spectrum of particles we observe.

> *"We were looking for the principles that determine the S-matrix. We found that those principles pointed to one-dimensional strings."*
>
> — Yoichiro Nambu

From Points to Strings

The dual models described particle interactions without using field theory. They naturally suggested that the underlying objects had length.

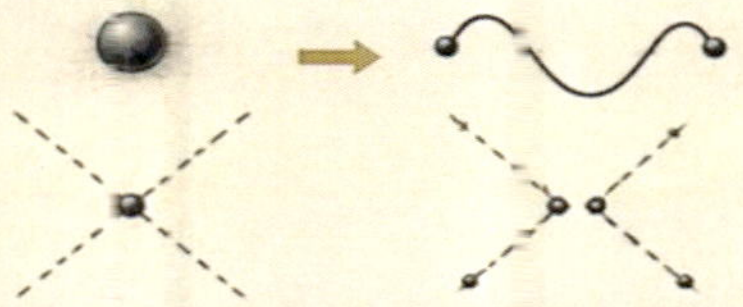

Vibrations Create Particles

Each vibrational mode of a string corresponds to a different particle.

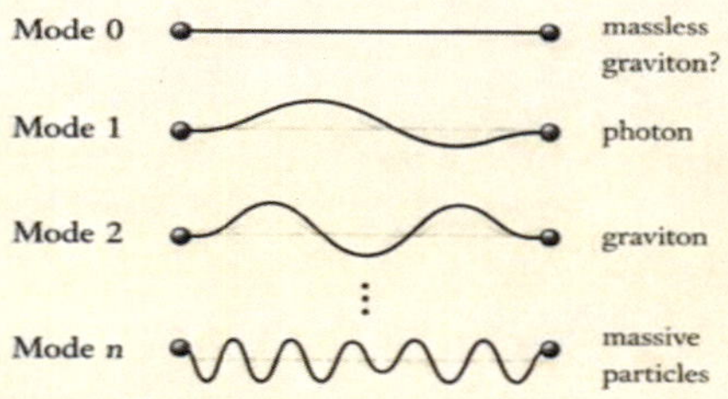

A Unifying Vision

Strings can, in principle, describe all fundamental particles and forces within a single framework.

One theory. Many particles.
All connected by strings.

Key Pioneers

Leonard Susskind · Holger Nielsen · Yoichiro Nambu

Their groundbreaking work revealed that strings, not points, might be the true constituents of nature.

The Big Idea

- ★ Replace point particles with one-dimensional strings.
- ★ Vibrations of strings explain the particles and their properties.
- ★ Interactions arise from strings spllitting and joining.
- ★ A promising path toward a unified theory of everything.

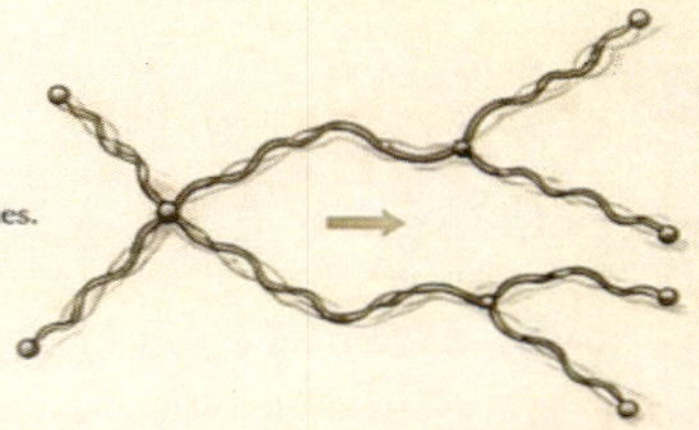

Milestones of the Early 1970s

1970
Dual Models
Dual resonance models describe strong interactions without field theory.

1971
Veneziano Amplitude
Revealed deep mathematical structure hinting at extended objects.

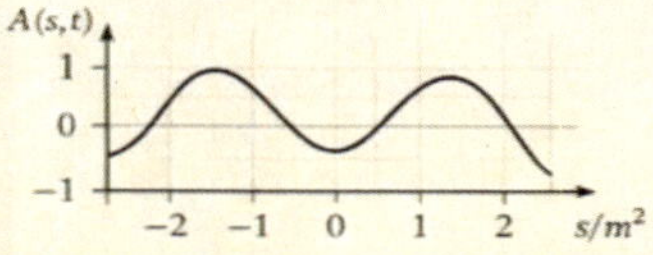

1972
Strings Emerge
Nambu, Nielsen, Susskind and others propose that the fundamental objects are one-dimensional strings.

Late 1970s
Toward a Theory
Initial excitement leads to a surge of ideas and the search for a consistent, unified string theory.

> *"The concept of the vibrating string was born from the quest to understand nature at its most fundamental level."*
>
> — Yoichiro Nambu

THE HISTORY OF

STRING THEORY

8

8. SYMMETRY AND BEAUTY

1970s

Guiding Principles

In the 1970s, physicists realized that symmetry is more than elegance—it is a guiding principle of nature.

String theory revealed that deep symmetries and consistency conditions could shape the theory and hint at a more fundamental structure of reality.

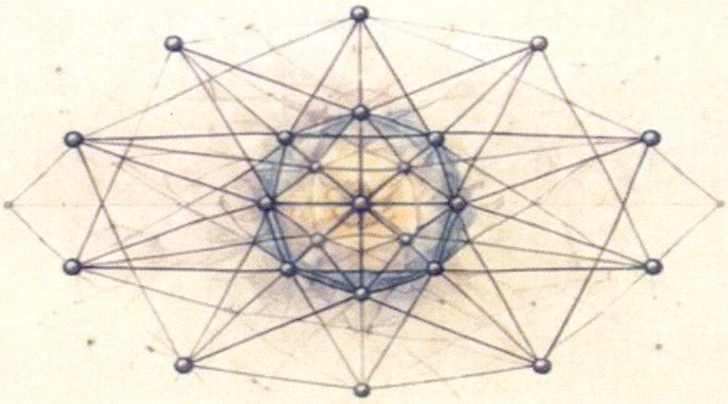

KEY IDEAS

★ Symmetries restrict what is possible.

★ Consistency conditions guide the form of the theory.

★ Beauty in mathematics often points to truth in nature.

PILLARS OF STRING THEORY

1. GAUGE SYMMETRIES

Forces arise from symmetries. Changing the viewpoint (a gauge transformation) does not change physical reality.

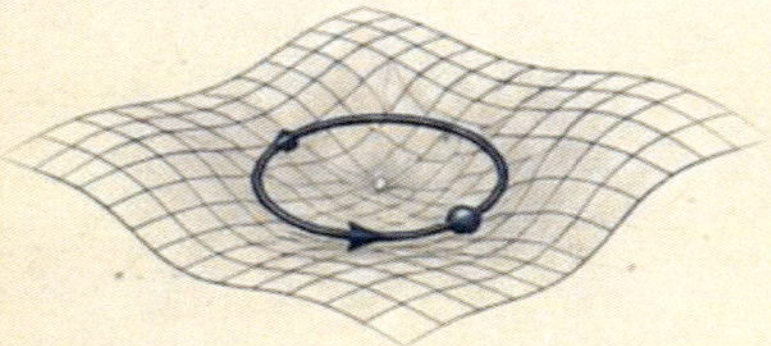

Different choices, same physics.

2. DUALITIES

Different theories can describe the same physics in surprising ways.

Different pictures, one underlying reality.

3. CONSISTENCY CONDITIONS

Anomaly cancellation and other consistency requirements eliminate inconsistent possibilities.

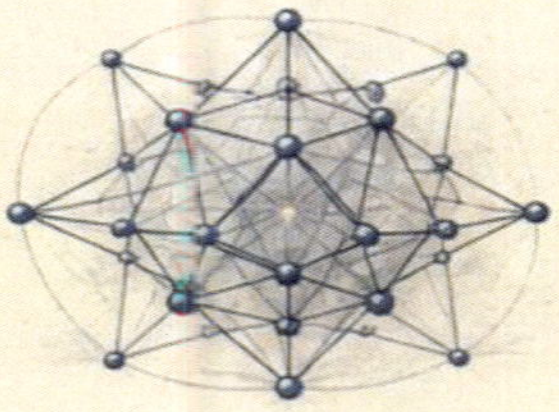

Consistency shapes the theory.

WHAT THESE PRINCIPLES REVEALED

★ Symmetry connected seemingly unrelated ideas.

★ Dualities showed that string theories are different facets of a single framework.

★ Mathematical beauty acted as a compass leading toward deeper truths.

"Beauty is the first test: there is no permanent place in the world for ugly mathematics."

— Freeman Dyson

A FOUNDATION FOR THE FUTURE

Early 1970s
The Search for Unity
Many approaches explored, seeking a deeper structure behind forces and particles.

Mid 1970s
Symmetry Revolution
Gauge symmetries and global symmetries become central to particle physics.

Late 1970s
The Power of Principles
Symmetry, duality, and consistency become the guiding stars.

1980s and Beyond
New Horizons
These principles open the door to supersymmetry, M-theory, and deeper ideas.

The Ongoing Quest
A Unified Truth
The journey continues, guided by symmetry, beauty, and curiosity.

"Symmetry is not a decoration of nature, but the very principle by which nature is constructed."

— Hermann Weyl

The History of String Theory

9

9. Superstrings

Mid 1970s

Supersymmetry Enters

In the mid 1970s, physicists discovered that combining supersymmetry with strings led to five consistent versions of a quantum theory of gravity.

These superstring theories opened the door to a deeper, more unified understanding of nature.

Different vibrations, same underlying superstring.

What is Supersymmetry?

Supersymmetry (SUSY) proposes a partner for every particle:

- ★ Bosons have fermion partners.
- ★ Fermions have boson partners.

This symmetry could solve deep mysteries and unify the forces of nature.

The Five Superstring Theories

All five theories are mathematically consistent. They differ in how their strings are oriented and what symmetries they possess.

1. Type I

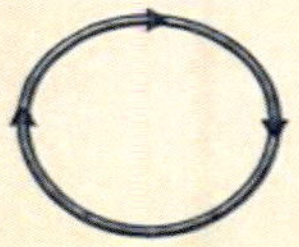

Closed strings with orientations and open strings with endpoints.

SO(32) gauge symmetry.

2. Type IIA

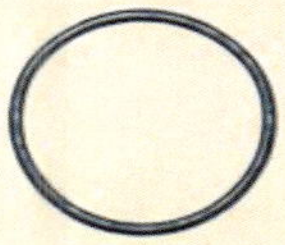

Closed strings only.

Non-chiral, non-oriented.

No tachyons.

3. Type IIB

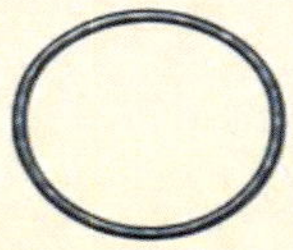

Closed strings only.

Chiral theory with high symmetry.

4. Heterotic SO(32)

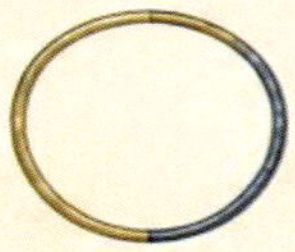

Left-moving modes in one sector, right-moving in another.

SO(32) gauge symmetry.

5. Heterotic $E_8 \times E_8$

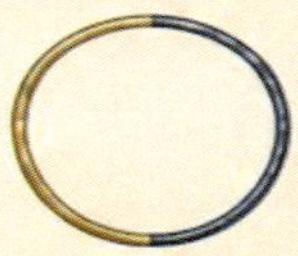

Left-moving modes in one sector, right-moving in another.

$E_8 \times E_8$ gauge symmetry.

Why Superstrings Matter

- ★ They include gravity naturally.
- ★ They remove inconsistencies that plague earlier models.
- ★ They suggest a unified framework for all fundamental forces.
- ★ They hint at a deeper structure behind space, time, and matter.

> *"Supersymmetry is the guiding star that led us to strings."*
>
> — John Schwarz

Key Developments and Impact

Mid 1970s
A New Ingredient

Supersymmetry becomes essential for consistency in string theory.

1976–1978
Consistent Theories

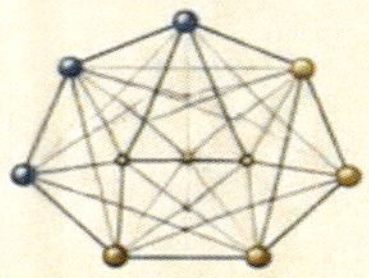

Five consistent superstring theories are discovered.

Late 1970s
Mathematical Power

Advanced mathematics and symmetry principles reveal hidden unity.

1980s
Toward Unification

Superstrings inspire new ideas and connections across physics.

The Legacy
A Lasting Impact

Superstrings remain a leading path toward a Theory of Everything.

> *"The hope is that strings and supersymmetry are the keys that will unlock the ultimate theory of nature."* — Brian Greene

THE HISTORY OF

STRING THEORY

10

10. THE FIVE SUPERSTRING THEORIES

LATE 1970s

Five consistent theories emerge, each with its own characteristics.

1. TYPE I

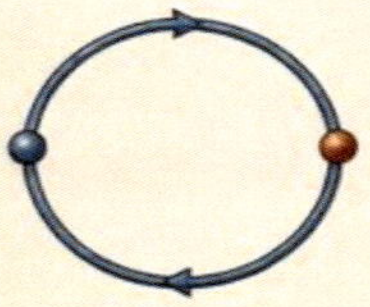

Open and closed strings.
Unoriented.
SO(32) gauge symmetry.
Includes D-branes.

2. TYPE IIA

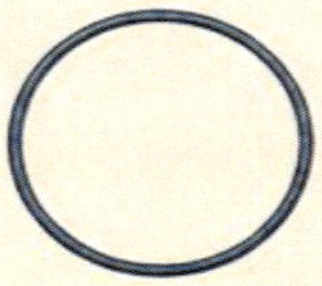

Closed strings only.
Non-chiral.
No tachyons.
Non-supersymmetric in 10D, supersymmetric when compactified.

3. TYPE IIB

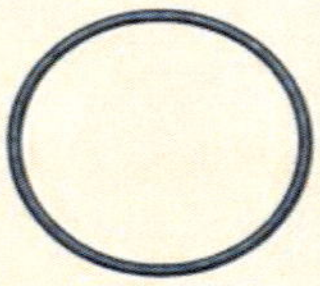

Closed strings only.
Chiral theory.

Maximal supersymmetry in 10 dimensions.

4. HETEROTIC SO(32)

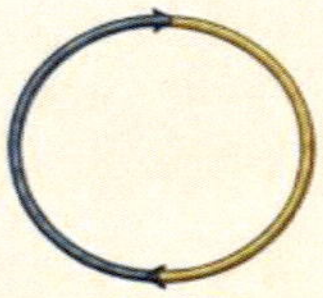

Left-moving modes: supersymmetric.
Right-moving modes: bosonic.
SO(32) gauge symmetry.

5. HETEROTIC $E_8 \times E_8$

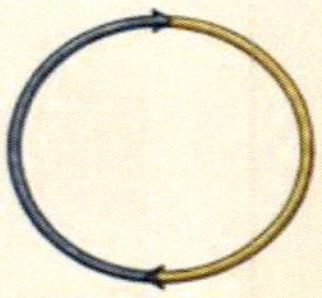

Left-moving modes: supersymmetric.
Right-moving modes: bosonic.
$E_8 \times E_8$ gauge symmetry.

> *"These five theories represent all the consistent ways to combine quantum mechanics, general relativity, and supersymmetry using strings."*
> — Michael Green, John Schwarz, and Edward Witten

COMMON FEATURES

- ★ All require ten space-time dimensions.
- ★ All include supersymmetry.
- ★ All are free of quantum anomalies.
- ★ All contain gravity (the graviton emerges naturally).
- ★ All describe the same physics at a fundamental level, though in different limits.

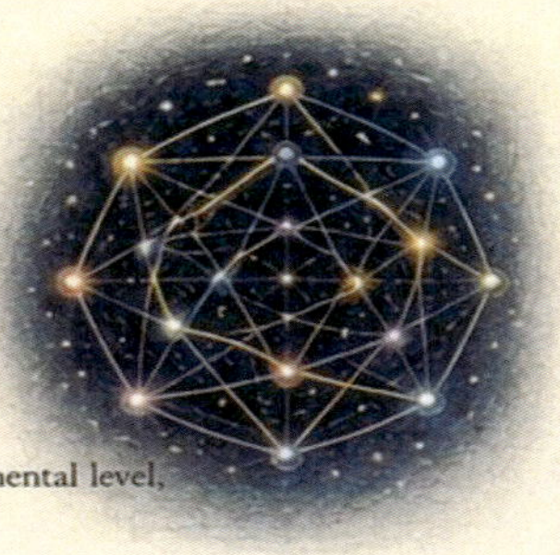

WHY FIVE?

String theory allowed for different ways to assign symmetries to the strings. Remarkably, these five possibilities are the only consistent ones. This hints at a deep mathematical structure underlying nature.

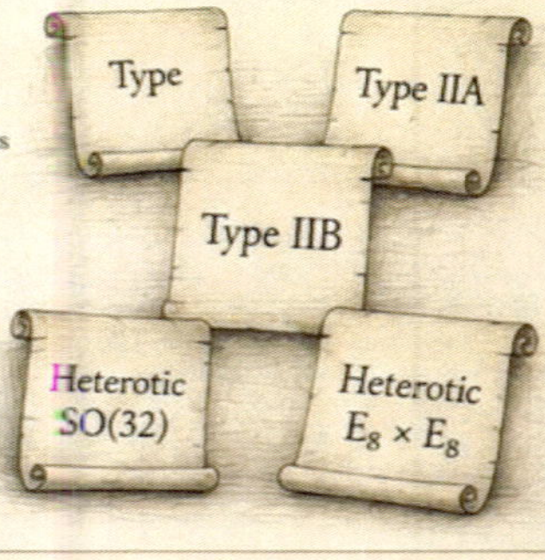

KEY MILESTONES

1974–1976
Supersymmetry Enters Physics

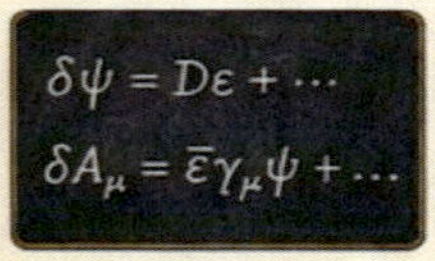

Supersymmetry provides a new symmetry relating bosons and fermions.

1976–1977
Anomalies Eliminated

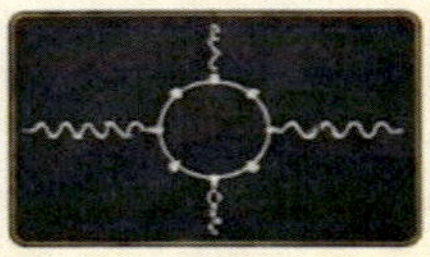

Anomaly cancellation conditions restrict possible string theories.

1978
New Consistent Theories

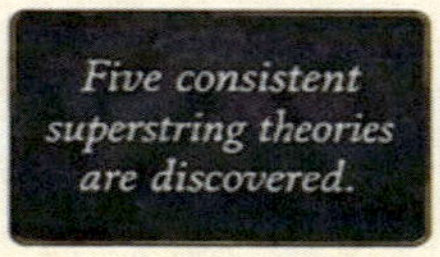

Different combinations of symmetries lead to five unique theories.

Late 1970s
Dualities and Unification

Dualities hint that these five theories may be different limits of one.

The Road Ahead
Toward a Unified Framework

The pursuit of a final, unified theory—M-theory—begins to take shape.

> *"Nature is not made of bricks, but of music.*
> *The five superstring theories are different songs of the same symphony."*
> — Edward Witten

11. M-Theory Conjecture

1990s

A Unifying Vision

By the early 1990s, five consistent superstring theories had been discovered. They seemed different, yet hints of a deeper unity kept emerging.

Physicists proposed that these theories might be different limits of a single, more fundamental framework: **M-theory.**

M-theory: a single theory underlying all consistent superstring theories.

> *"We are convinced that all superstring theories are different limits of a single underlying theory."*
>
> — Edward Witten
> (1995, Reflecting on the M-Theory Revolution)

Why M-Theory?

Five Theories, One Origin?

The five superstring theories seem distinct, but dualities revealed deep connections among them.

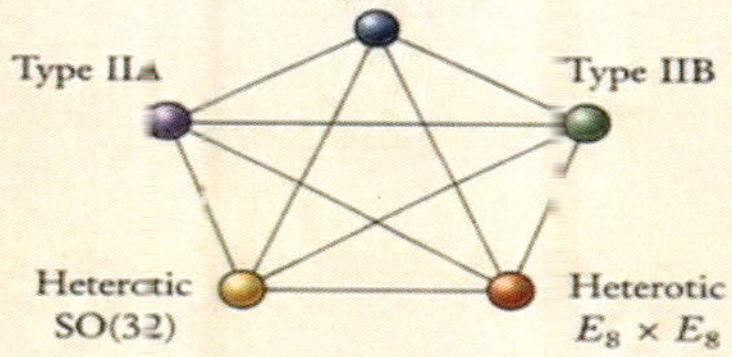

M-theory is the "hub" from which all five arise.

Beyond Strings

M-theory is not just about strings—it includes higher-dimensional objects called **branes.**

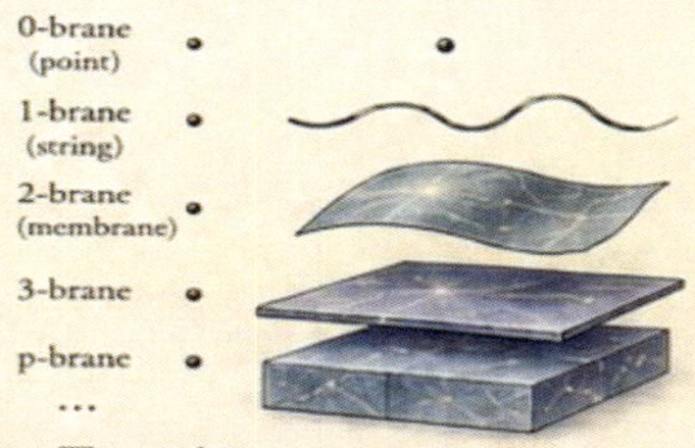

These objects can exist in many dimensions and interact in rich ways.

11 Dimensions

M-theory naturally lives in eleven dimensions:

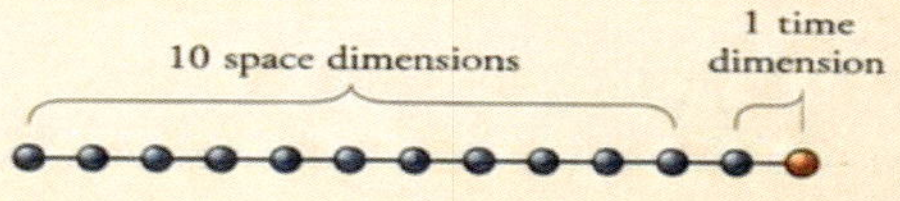

The extra dimensions are compactified or hidden from everyday experience, explaining why we see only 4.

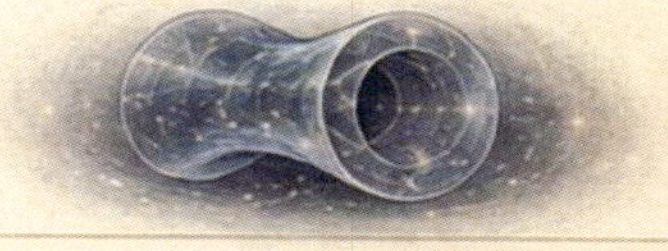

Dualities as Clues

Various dualities—like T-duality, S-duality, and U-duality—suggest that what looks different at low energies can be unified at a deeper level.

T-Duality

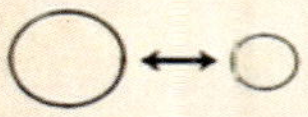

Large and small radii can be equivalent.

S-Duality

Strong coupling in one theory = weak coupling in another.

U-Duality

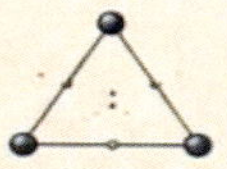

Unifies multiple types of dualities.

What M-Theory Promises

- ★ A single framework containing all known consistent superstring theories.
- ★ A quantum theory of gravity.
- ★ A deeper understanding of space, time, and the fundamental structure of the universe.
- ★ The possibility of explaining everything—from black holes to cosmology—in one theory.

Key Developments of the 1990s

1990
Dualities Discovered
T-duality and S-duality reveal unexpected equivalences.

1991–1993
Five Superstring Theories
Five consistent superstring theories are found.

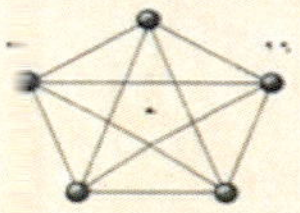

1995
M-Theory Hypothesis
Edward Witten and others propose M-theory as the unifying framework.

1996–1999
Branes and Beyond
D-branes and extended objects become central to the theory.

Toward the Future
A Deep Quest Continues
Mathematics and physics advance together in the pursuit of ultimate unity.

> *"M-theory is our best hope for a final theory. We may not know its full form yet, but we are learning how the pieces fit together."* — Cumrun Vafa

12. DUALITIES

1990s

Different Views, Same Physics

In the 1990s, physicists uncovered powerful dualities—deep equivalences between theories that looked very different.

Dualities revealed that nature doesn't have just one description. There are many views of the same underlying reality.

Different pictures, one underlying truth.

> *"Dualities are the Rosetta Stone of string theory. They tell us that different theories are just different languages describing the same physics."*
>
> — Juan Maldacena

MAJOR DUALITIES DISCOVERED

T-DUALITY

Small and Large Are the Same

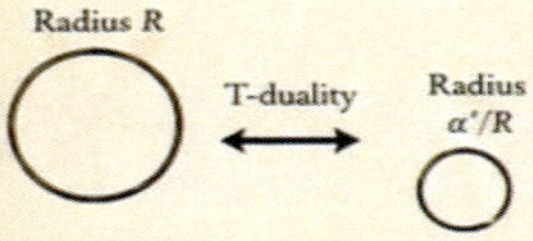

A string moving on a circle of radius R is equivalent to a string on a circle of radius α'/R. Physics remains unchanged.

S-DUALITY

Strong Becomes Weak

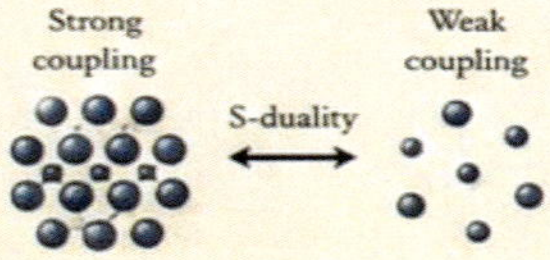

A strongly coupled theory can be equivalent to a weakly coupled one.

It lets us study the unstudyable.

U-DUALITY

A Larger Web of Equivalences

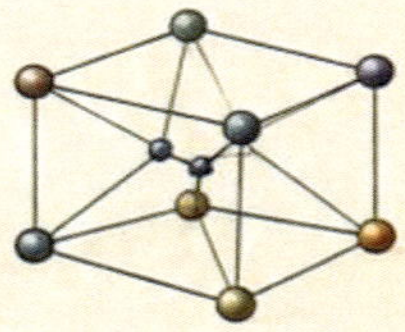

Combines T- and S-dualities into a larger symmetry group.

Many theories are connected by U-duality.

ADS/CFT DUALITY

Gravity = Quantum Field Theory?

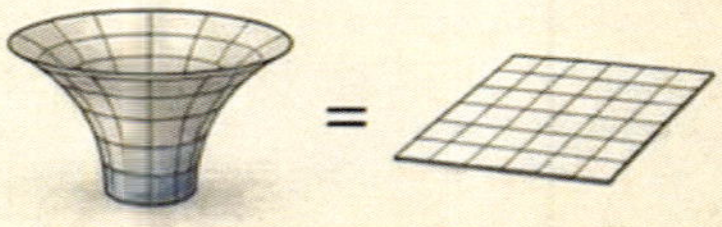

Gravity in a curved (AdS) space

Quantum field theory on the boundary

Proposed by Maldacena (1997). A deep link between gravity and quantum theory.

WHAT DUALITIES TEACH US

- ★ No description is the "one and only" way.
- ★ What looks different can be the same.
- ★ New tools for solving hard problems.
- ★ Clues toward a deeper, more unified theory.
- ★ Hints that the universe is more interconnected than we imagined.

IMPACT ON PHYSICS

- ★ Helped reveal the unity behind all five superstring theories.
- ★ Opened new ways to study black holes and quantum gravity.
- ★ Led to breakthroughs in mathematics and geometry.
- ★ Inspired powerful ideas in condensed matter and quantum systems.

A DECADE OF DISCOVERY

1990–1991
First Dualities
T-duality and S-duality are discovered in string theory.

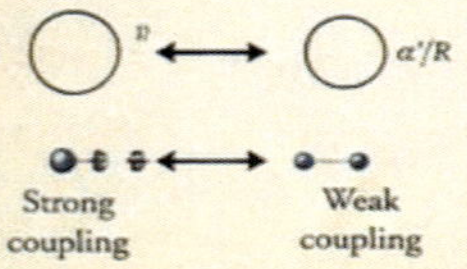

1992–1994
U-Duality Emerges
A larger symmetry connects many string theories in new ways.

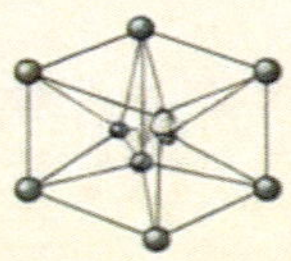

1995–1996
Black Holes and Beyond
Dualities provide new insights into black holes, entropy, and quantum gravity.

1997
AdS/CFT Conjecture
Maldacena proposes a bold equivalence between gravity in bulk space and a field theory on the boundary.

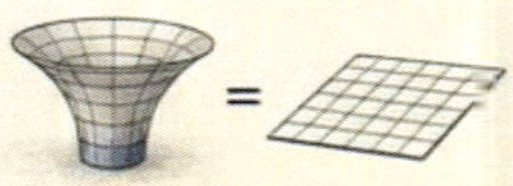

Late 1990s
A Web of Unity
Dualities weave the theories together into a beautiful, interconnected whole.

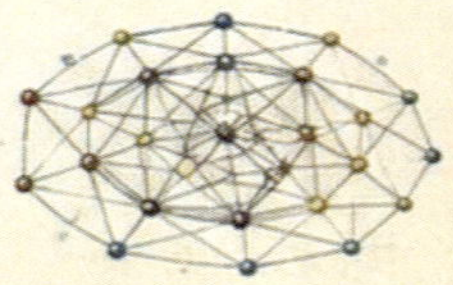

> *"Dualities show us that nature is a master of disguise. What seems different may be the same song sung in another key."*
>
> — Nathan Seiberg

THE HISTORY OF STRING THEORY

13. BRANES

1990s

Higher-Dimensional Objects

In the 1990s, physicists realized that strings are not the only extended objects in string theory.

Branes (short for "membranes") generalize strings to higher dimensions. They opened new ways to understand space, time, and the universe.

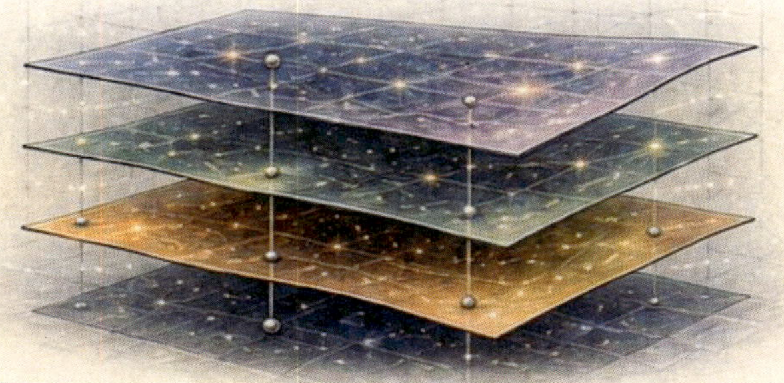

Branes: higher-dimensional objects on which strings can end or move.

KEY IDEA

Just as particles are points and strings are one-dimensional, branes can have two, three, or more dimensions.

They can exist in higher-dimensional space and interact in rich and surprising ways.

TYPES OF BRANES

0-BRANE (POINT PARTICLE)

$p = 0$

A particle with no spacial extension.

1-BRANE (STRING)

$p = 1$

A string with one spatial dimension.

2-BRANE (MEMBRANE)

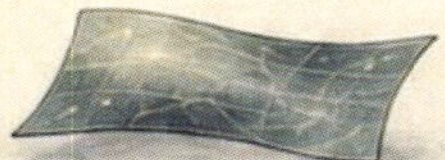

$p = 2$

A membrane with two spatial dimensions.

3-BRANE

$p = 3$

A three-dimensional object.

***p*-BRANE**

p = any

Branes can have any number of dimensions.

Branes generalize strings ($p = 1$) to p spatial dimensions. They can exist in higher-dimensional spaces.

OPEN AND CLOSED STRINGS ON BRANES

OPEN STRINGS

Open strings can end on branes.

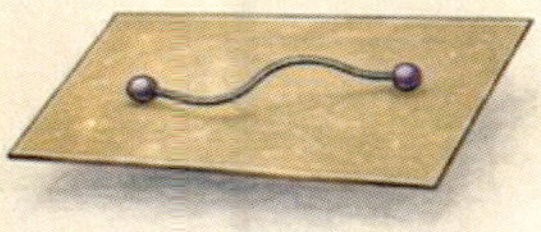

CLOSED STRINGS

Closed strings can move in the bulk (the higher-dimensional space).

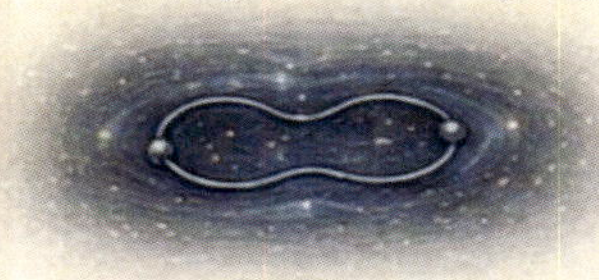

Branes provide "homes" for open strings and gauge forces.

WHY BRANES MATTER

★ They provide new ways to unify forces and particles.

★ They led to breakthroughs in black hole physics.

★ They connect string theory to quantum gravity and cosmology.

★ They inspired powerful ideas in mathematics, including derived categories and topology.

★ They are central to modern developments in theoretical physics.

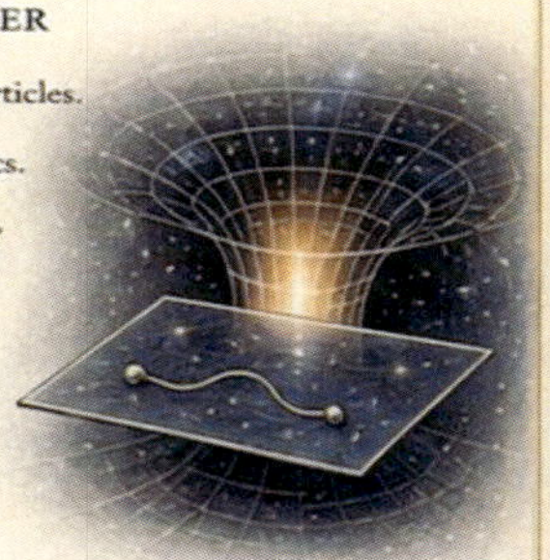

MAJOR ADVANCES IN THE 1990s

1991
D-Branes Discovered

D-branes are dynamical objects where open strings can end.

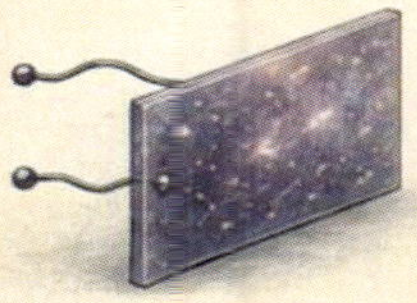

1995
Branes and Dualities

Branes provided new insight into dualities and strong/weak coupling.

1996
Branes and Black Holes

Brane ideas helped count black hole microstates and explain entropy.

1996–1997
Brane-World Scenarios

Our universe might be a 3-brane moving in a higher-dimensional space.

Late 1990s
M-Theory and Branes

Branes became central to M-theory and the search for a unified framework.

"Branes opened the door to new dimensions of understanding. They showed us that the universe can be far richer than we imagined."

— Joseph Polchinski

The History of String Theory

14. String Theory and Black Holes

1990s

When Gravity Meets Strings

The 1990s brought a remarkable convergence: string theory—originally developed without black holes in mind—began to shed light on the deepest mysteries of gravity.

Strings offered new mathematical tools to study black holes, and black holes revealed new tests for string theory.

Black holes became a powerful laboratory for ideas about quantum gravity.

> *"Black holes are where gravitational collapse meets quantum theory. Strings give us a new set of tools to explore this frontier."*
>
> — Juan Maldacena

Key Breakthroughs

1. Black Hole Entropy from Strings

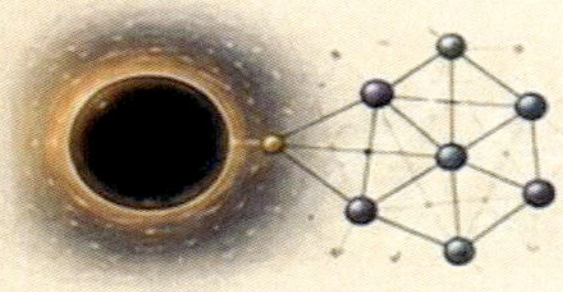

Strominger and Vafa (1996) showed that counting microstates of certain black holes in string theory reproduces the Bekenstein–Hawking entropy. A triumph for quantum gravity.

2. D-Branes as Black Hole Building Blocks

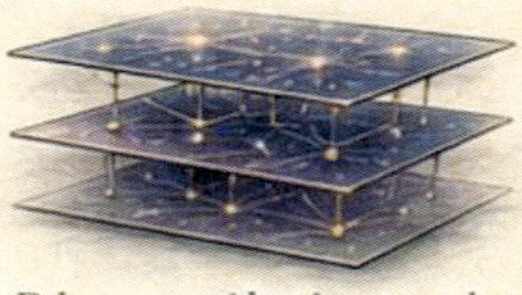

D-branes provide microstates that explain black hole entropy and dynamics. They revealed a deep connection between geometry and quantum states.

3. AdS/CFT Correspondence (1997)

Maldacena's conjecture related gravity in Anti–de Sitter (AdS) space to a quantum field theory on its boundary. A revolutionary holographic duality.

4. Black Hole Information and Holography

String theory suggested that information is not lost in black holes—it is encoded on the boundary. The holographic principle was born.

Strings Resolve Singularities

In classical general relativity, black holes contain singularities—points of infinite density. String theory smooths them out.

- ★ No infinities.
- ★ New physics at tiny scales.
- ★ Extended objects change everything.

Why It Matters

- ★ Black holes became a testing ground for string theory.
- ★ D-branes unified gravity, gauge theories, and geometry.
- ★ Holography suggested that spacetime itself may emerge from deeper quantum degrees of freedom.
- ★ These ideas continue to shape modern physics and the search for quantum gravity.

A 1990s Timeline

1991
D-Branes Introduced
Polchinski introduced D-branes, extending string theory's symmetry and power.

1992–1993
Dualities Connect Theories
Dualities revealed deep links between seemingly different string theories.

1996
Black Hole Entropy Explained
Strominger and Vafa matched microscopic string states to black hole entropy.

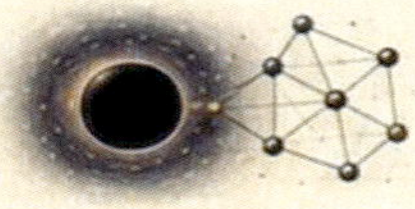

1997
AdS/CFT Conjecture
Maldacena proposed the holographic duality between gravity and quantum fields.

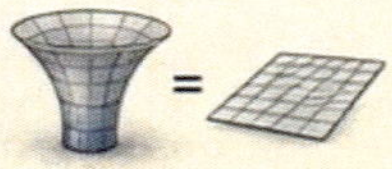

1998–1999
String Theory Matures
Black holes, branes, and holography reshape our understanding of the universe.

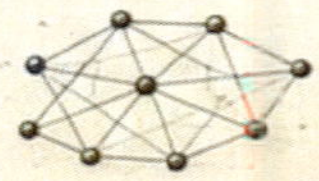

Toward the 21st Century
A New Framework Emerges
String theory becomes a leading candidate for a theory of everything.

> *"Black holes taught us that the universe keeps its secrets on the boundary. Strings gave us the language to read them."*
>
> — Leonard Susskind

The History of String Theory

15. The AdS/CFT Correspondence

1997

A Powerful Conjecture

In 1997, Juan Maldacena proposed a bold and surprising duality: a theory of gravity in a space with one extra dimension is equivalent to a quantum field theory with no gravity on the boundary of that space.

This idea—known as the AdS/CFT correspondence—became one of the most profound insights in theoretical physics.

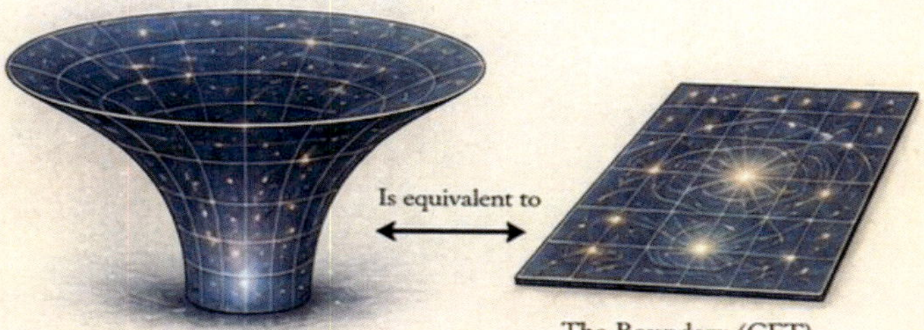

The Bulk (AdS Space)
A space with gravity.

The Boundary (CFT)
A quantum field theory with no gravity.

> *"The equivalence was like a hologram: all the information in the bulk is encoded on the boundary."*
>
> — Juan Maldacena (1997)

How it Works

1. A Space with Gravity

Consider Anti-de Sitter (AdS) space, a curved spacetime with a negative cosmological constant.

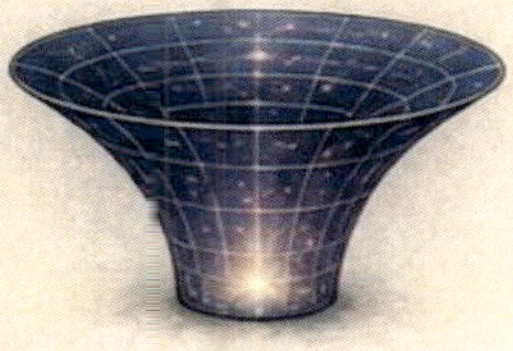

Gravity lives in the bulk.

2. A Conformal Field Theory

On the boundary of AdS space lives a conformal field theory (CFT) with no gravity.

Physics without gravity.

3. A Surprising Duality

Maldacena proposed these two seemingly different theories are exactly equivalent.

Two sides of the same coin.

4. A Holographic Universe

All information in the bulk—gravity, black holes, strings—has a complete description on the boundary.

The universe as a hologram.

What it Connects

Gravity (Strongly Coupled)		Quantum Fields (Weakly Coupled)
AdS Space (Bulk)	⟷ Duality	CFT (Boundary)

A difficult problem with gravity in the bulk can be easier to solve as a quantum field theory on the boundary—and vice versa.

Why it Matters

- ★ Provides a non-perturbative definition of quantum gravity in AdS space.
- ★ Offers new tools to study strongly coupled systems in particle physics and condensed matter.
- ★ Has inspired discoveries about black holes, quantum information, entanglement, and spacetime geometry.
- ★ Suggested that spacetime is not fundamental—it emerges from quantum entanglement.

Key Developments after 1997

1998–2000
Testing the Duality
Many checks confirmed aspects of the correspondence.

Agreement on both sides grew stronger.

2001–2005
Beyond AdS_5/CFT_4
Generalizations to other dimensions and backgrounds were developed.

A broader holographic framework emerged.

2006–2010
Applications Expand
New connections to quark-gluon plasma, critical phenomena, and condensed matter systems.

Holography goes beyond high-energy physics.

2011–2015
Quantum Information
Entanglement entropy and information geometry deepened the link between information and spacetime.

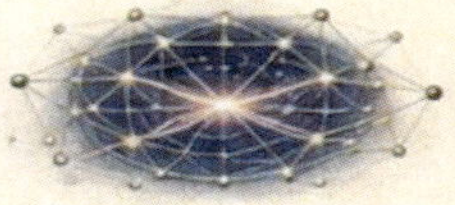

"ER=EPR" and related ideas spark new insights.

2016–Present
A Living Correspondence
Holography continues to inspire new mathematics, experiments, and unifying ideas.

The holographic universe keeps revealing itself.

> *"The AdS/CFT correspondence was a turning point. It taught us that gravity and quantum physics are not enemies, but two views of the same underlying reality."*
>
> — Leonard Susskind

16. LANDSCAPE (2000s)

A Multitude of Vacua

Advances in string theory revealed an astounding possibility: the theory admits an enormous number of consistent solutions—vacua—each corresponding to a different universe with its own physical laws.

This "landscape" of possibilities changed how physicists think about the fundamental nature of reality.

A vast landscape of possible universes.

> *"String theory may not predict a unique universe. Instead, it may describe a vast landscape of possibilities."*
>
> — Leonard Susskind (2003)

KEY INGREDIENTS

1. EXTRA DIMENSIONS

Different shapes and sizes of the extra dimensions lead to different physical laws in four dimensions.

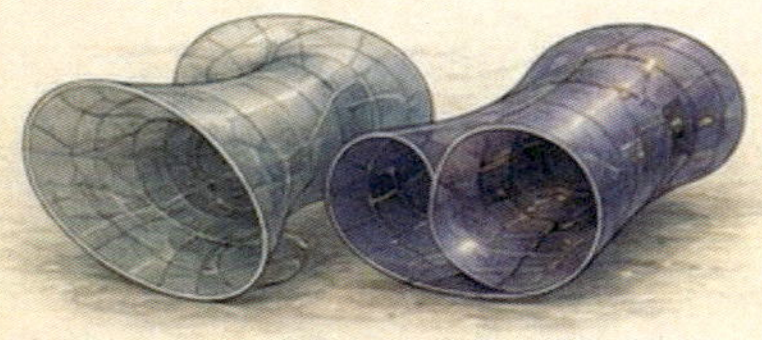

Many possible shapes (e.g., Calabi–Yau manifolds) for the hidden dimensions.

2. FLUXES

Fields can thread through the extra dimensions in many different ways.

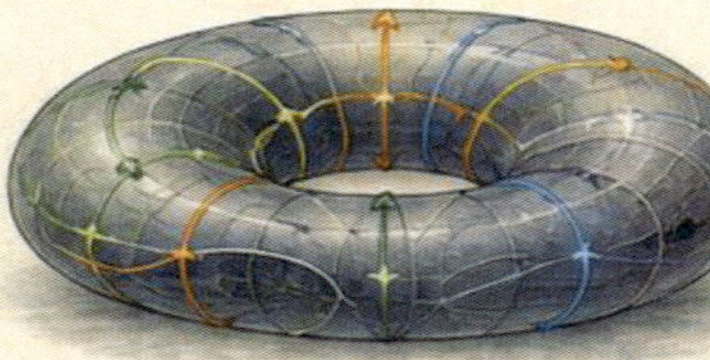

Different flux choices stabilize the geometry and lead to different vacua.

3. BRANES AND ORIENTIFOLDS

Branes and orientifolds can be arranged in many ways, adding even more variety to the solutions.

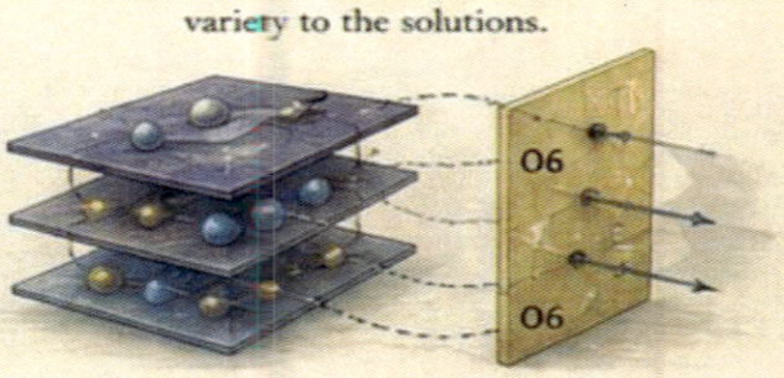

Many configurations produce different gauge groups and particle spectra.

THE LANDSCAPE

- ★ Estimates suggest 10^{100} to 10^{500} or more possible vacua.
- ★ Each vacuum has its own values for constants, particles, and forces.
- ★ Our universe is just one tiny possibility among many.
- ★ Explaining why our vacuum has the properties it does is one of the greatest challenges.

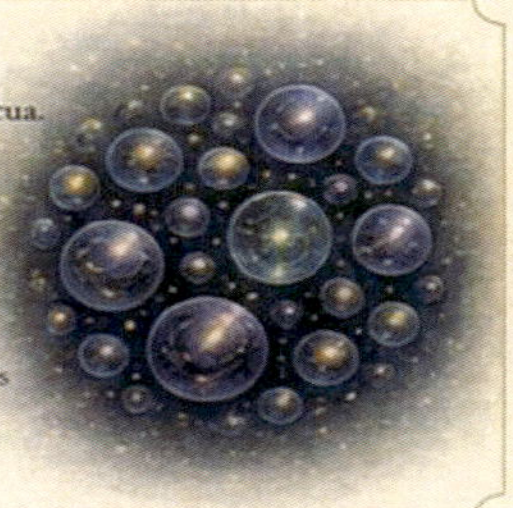

IMPLICATIONS

- ★ Predicting unique values becomes extremely difficult.
- ★ Anthropic reasoning may be needed: only some vacua allow observers.
- ★ New ideas in cosmology, statistics, and probability become crucial.
- ★ The landscape inspires both excitement and deep debate.

DEVELOPMENTS IN THE 2000s

2000–2002
Flux Compactifications
Turning on fluxes in extra dimensions stabilizes shapes and creates many vacua.

2003–2004
Branes in the Mix
Branes and orientifolds add new ingredients and possibilities.

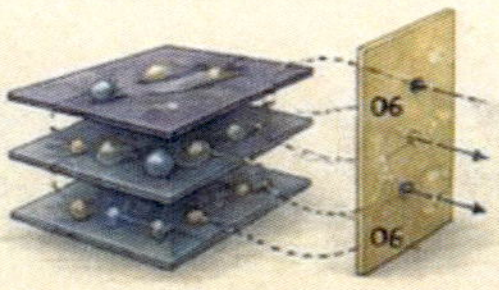

2005
Counting Vacua
Methods for counting vacua suggest a number that is enormous—perhaps enormous beyond measure.

2006–2008
Statistical Approaches
Physicists begin using probability and statistics to study the landscape.

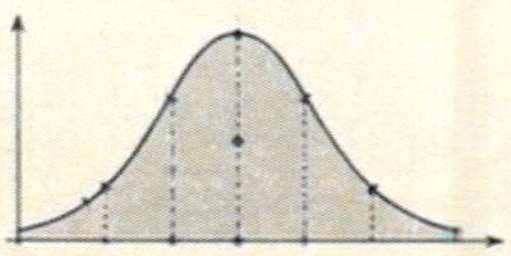

Late 2000s
New Questions
How do we choose a vacuum? What is the measure? What is the role of cosmology?

> *We may live in a vast multiverse of possibilities.*
> *String theory gives us a map of the landscape, but not yet the key to our location.*
>
> — Cumrun Vafa

THE HISTORY OF STRING THEORY

17. PHENOMENOLOGY (2000s)

Connecting to the Real World

As string theory matured, a central goal became clear: connect the beauty of the mathematics to the messy world of experiments.

Phenomenology builds bridges between the string landscape and observable physics—seeking signals we might one day detect.

From high-energy colliders to cosmic observations, we look for clues that string theory is real.

> *"Phenomenology is where ideas meet data. It is the compass that guides string theory toward physical reality."*
>
> — Nathan Seiberg (2006)

KEY APPROACHES

1. PARTICLE PHYSICS MODELS

Building models from string theory that can reproduce the Standard Model of particle physics.

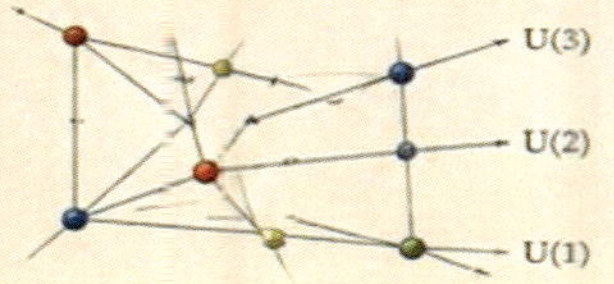

Intersecting branes can give rise to gauge groups and matter fields.

2. SUPERSYMMETRY

SUSY remains a key feature in many string constructions, helping to stabilize the hierarchy.

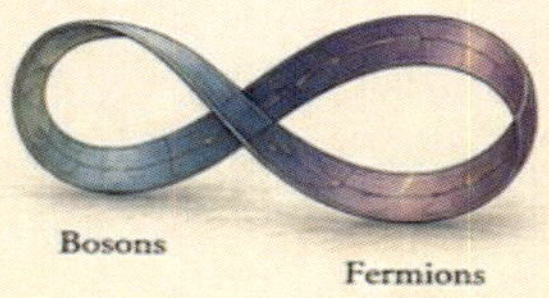

Supersymmetry relates bosons and fermions, reducing quantum corrections.

3. DARK MATTER CANDIDATES

String theory offers natural candidates for dark matter, often from hidden sectors or extra dimensions.

Hidden sector particles could make up the dark matter in our universe.

4. MODULI AND THE COSMOLOGICAL CONSTANT

Moduli fields shape the vacuum and may play a role in the small value of the cosmological constant.

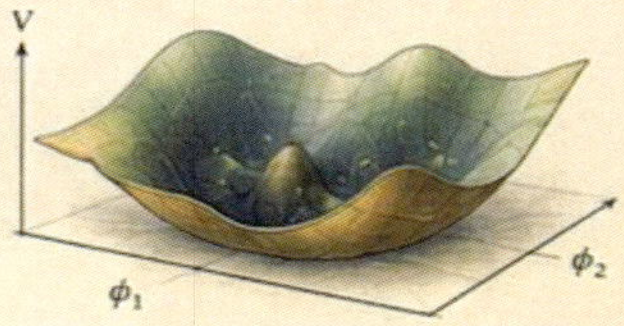

Stabilizing moduli is essential for predictive models and cosmology.

BRIDGING THEORY AND OBSERVATION

- ★ Look for deviations from the Standard Model at colliders.
- ★ Search for supersymmetry and extra dimensions.
- ★ Use cosmology to probe the early universe.
- ★ Study rare processes and precision measurements.
- ★ Develop new observables and experimental strategies.

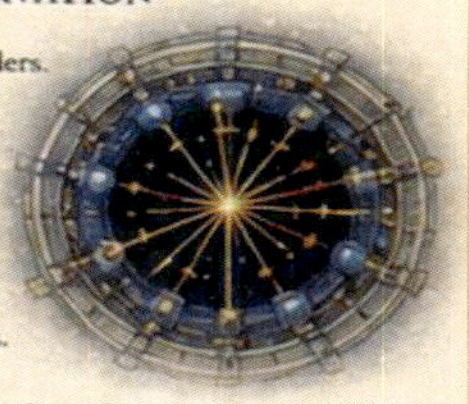

The Large Hadron Collider (LHC) explores physics at the highest energies.

CHALLENGES

- ★ The string scale may be far beyond current collider energies.
- ★ The landscape is vast—pinning down the right vacuum is difficult.
- ★ Many models, few unique predictions.
- ★ Connecting low-energy physics to high-energy geometry is complex.
- ★ New ideas and tools are still needed.

The path from mathematics to measurement is long and winding.

MILESTONES IN PHENOMENOLOGY

2000–2002
Realistic String Models
Intersecting brane models begin to reproduce key features of the Standard Model.

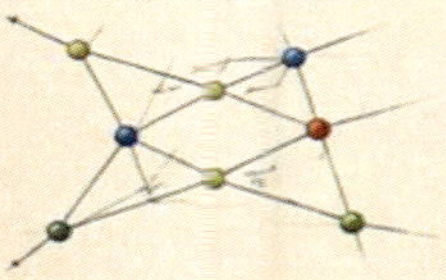

2003–2005
SUSY and Stabilization
Progress in moduli stabilization and the understanding of supersymmetry breaking.

2006–2008
Dark Matter and Hidden Sectors
Hidden sector candidates and axion-like particles emerge from string constructions.

2009–2011
Precision and Indirect Searches
Improved experimental limits guide model building and suggest new signatures.

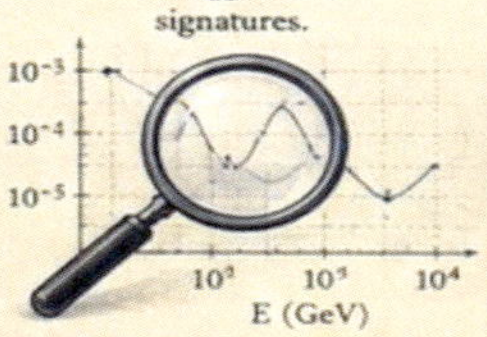

2012–2010s
Toward Discoverability
New ideas link string theory to cosmology, gravitational waves, and future experiments.

> *"Phenomenology is the art of turning mathematical possibility into experimental testability."* — Nima Arkani-Hamed

THE HISTORY OF STRING THEORY

18. MATHEMATICS AND STRING THEORY (2000s)

A Beautiful Partnership

As string theory matured, it became clear that mathematics was not just a tool—it was a partner in discovery.

Deep structures in geometry, topology, and algebraic mathematics found natural homes in string theory. In return, string theory inspired new mathematics.

Calabi–Yau manifolds: complex shapes that may hide extra dimensions.

> *"String theory has been a catalyst for some of the most exciting developments in mathematics in the past several decades."*
>
> — Cumrun Vafa (2007)

MATHEMATICAL AREAS INFLUENCED

1. GEOMETRY

Extra dimensions are often shaped like Calabi–Yau manifolds and other complex geometric spaces.

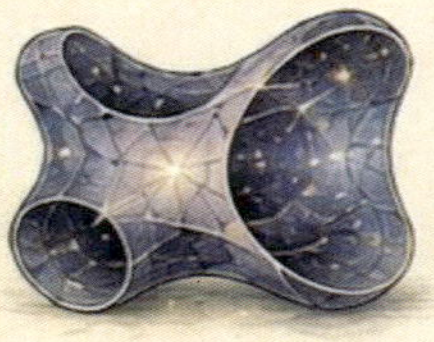

Geometry describes the shape of the universe at the smallest scales.

2. TOPOLOGY

Topology studies properties that remain unchanged under continuous deformation.

String theory uses topology to classify consistent string backgrounds.

3. ALGEBRA AND REPRESENTATION THEORY

Symmetries in string theory are encoded in algebraic structures and their representations.

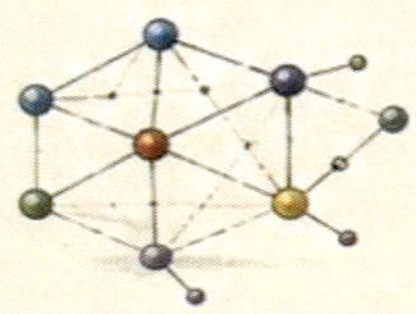

These structures describe particles, forces, and dualities in precise ways.

4. NUMBER THEORY

Surprising connections emerged between strings and deep questions in number theory.

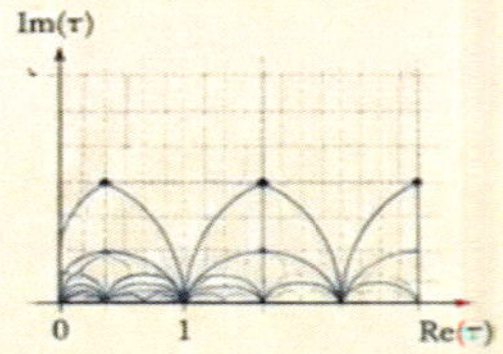

Objects like modular forms appear in counting states and black hole entropy.

5. CATEGORY THEORY

Modern mathematical physics uses category theory to understand structures and dualities at a high level.

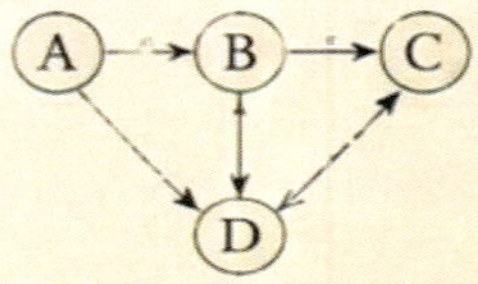

It provides a unifying language for many areas of modern physics.

MIRACULOUS MATHEMATICS

- ★ New invariants and formulas emerged from string theory problems.
- ★ Enumerative geometry grew from counting curves in Calabi–Yau spaces.
- ★ Knot theory, derived categories, and motivic ideas found new motivation.
- ★ Many "pure" mathematical ideas first appeared in string theory research.

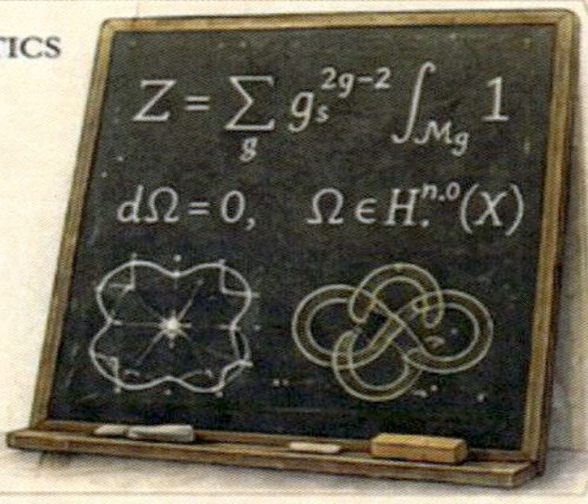

WHY IT MATTERS

- ★ Mathematics provides the language to describe string theory precisely.
- ★ New mathematical tools help solve hard physical problems.
- ★ The partnership deepens our understanding of both mathematics and the universe.
- ★ Beauty and consistency guide us toward deeper truths.

MAJOR DEVELOPMENTS IN THE 2000s

2000–2002
Mirror Symmetry and Beyond

Mirror symmetry connects pairs of Calabi–Yau manifolds in deep and surprising ways.

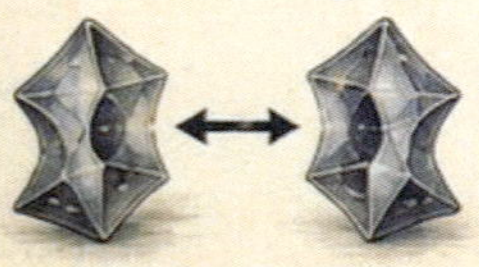

A duality with powerful mathematical consequences.

2003–2005
Topological Strings

New "topological" versions of string theory led to invariants and exact mathematical results.

Solving counting problems with precision.

2006–2008
Geometric Transitions

Transitions between different geometries explained how spacetime itself can change in string theory.

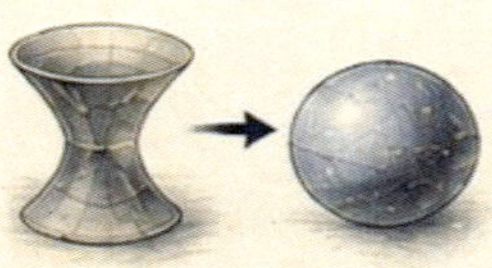

Geometry is dynamic at the deepest level.

2009–2011
Homological Methods

Tools from homological algebra and derived categories became essential in many areas.

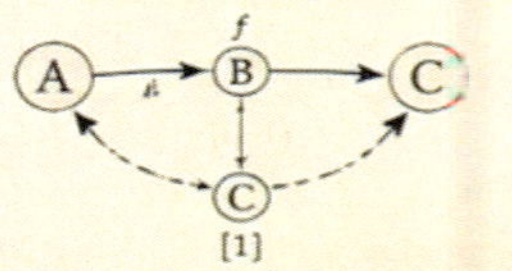

A new language for dualities and equivalences.

2012–Late 2000s
Mathematical Physics Comes of Age

String theory and mathematics entered a golden era of cross-fertilization.

Stronger together—advancing knowledge on both sides.

> *"The most incomprehensible thing about the universe is that it is comprehensible."*
>
> — Albert Einstein

19. String Theory Today (2000s)

An Ongoing Journey

String theory has traveled a remarkable path—from a daring idea about hadrons to a bold framework that could unite all fundamental forces, including gravity.

Today it stands at the frontier of physics, guiding research, inspiring debate, and opening windows to realities we are only beginning to understand.

> *"We are each travellers in the wilderness of wonder... Our goal is not to arrive, but to understand."*
> — David Gross (2005)

Active Frontiers of Research

1. Quantum Gravity

String theory aims to consistently quantize gravity and reveal the true nature of space and time at the smallest scales.

2. The Landscape and Vacua

The theory allows a vast number of possible vacuum solutions—each with different physical laws. Why our universe?

3. Black Holes and Holography

Holographic dualities connect gravitational systems to quantum theories, shedding light on black holes and information.

4. Phenomenology and Cosmology

Connecting string theory to the real world: particle physics, inflation, dark matter, and the early universe.

5. Mathematics and Structure

Deep mathematics—geometry, category theory, number theory—continues to reveal hidden structures in the theory.

Achievements So Far

- ★ Five consistent superstring theories discovered.
- ★ Dualities revealed deep, unexpected equivalences.
- ★ Holographic principle transformed our view of spacetime.
- ★ Connections to black holes, quantum information, and strongly coupled systems.
- ★ Powerful new mathematics discovered along the way.

Challenges Ahead

- ★ No direct experimental evidence—yet.
- ★ The landscape may be too vast to predict uniquely.
- ★ Finding the right vacuum for our universe.
- ★ Developing new tools to make testable predictions.
- ★ Understanding time, causality, and emergence.

Recent Developments and Highlights

2000–2002
Revisiting Dualities
New dualities, including AdS/CFT, reshape our understanding of quantum gravity.

2003–2005
Flux Compactifications
Fluxes stabilize shapes of extra dimensions and lead to a richer set of possible vacua.

2006–2008
Gauge/Gravity Progress
Gauge/gravity dualities illuminate strongly coupled systems in new ways.

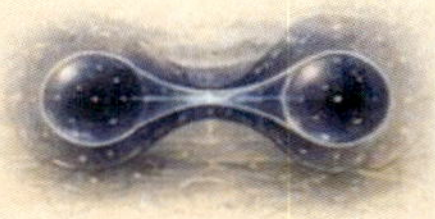

2009–2011
Cosmology Connections
String-inspired models tackle inflation, cosmic strings, and early universe physics.

2012–2016
New Mathematical Tools
Advances in geometry, topology, and categories deepen the theoretical foundation.

2017–Present
Ongoing Exploration
New ideas, new methods, and new questions keep the journey alive and full of promise.

> *"String theory is a journey toward a deeper truth about the universe. Each answer we find reveals more questions—and the adventure continues."*
> — Cumrun Vafa

THE HISTORY OF STRING THEORY

20. THE FUTURE (2010s AND BEYOND)

Beyond the Horizon

The journey of string theory continues. Though we have not yet reached a final theory of everything, the path ahead is rich with possibilities.

New ideas, technologies, and collaborations may one day reveal the profound truths that lie beyond the horizon.

"The future is not something we enter. The future is something we create."

— Leonard Nimoy (1931–2015)

PROMISING DIRECTIONS

1. THE SWAMPLAND PROGRAM

Exploring the vast landscape of string vacua to find principles that explain why our universe has the properties it does.

2. QUANTUM GRAVITY AND GEOMETRY

Deepening our understanding of spacetime, singularities, and the emergent nature of geometry from quantum entities.

3. HOLOGRAPHY AND EMERGENCE

Discovering how lower-dimensional theories may encode higher-dimensional realities through holographic dualities.

4. MATHEMATICS INSPIRING PHYSICS

New mathematical structures—from derived algebraic geometry to infinity categories—may unlock new physical insights.

5. CONNECTIONS TO OBSERVABLES

Bridging the gap between theory and experiment: could cosmic signals, gravitational waves, or black holes hold the key?

GRAND CHALLENGES

- ★ No direct experimental confirmation—yet.
- ★ The landscape may be too vast to search completely.
- ★ Understanding time, causality, and the arrow of time.
- ★ Incorporating quantum mechanics and gravity fully.
- ★ Finding a principle of uniqueness or necessity.
- ★ Explaining the origin of the universe itself.

WHY WE PURSUE THE UNKNOWN

- ★ To understand the deepest laws of nature.
- ★ To answer questions about where we came from and where we are going.
- ★ To unite the beauty of mathematics with the reality of the universe.
- ★ To inspire future generations to explore, question, and dream.
- ★ Because curiosity is the spark that lights the path forward.

THE JOURNEY CONTINUES

2010–2012
Refining Ideas

Progress in holography, black hole microstates, and mathematical frameworks.

2013–2014
Swampland and Constraints

New criteria emerge to distinguish viable theories from the vast landscape.

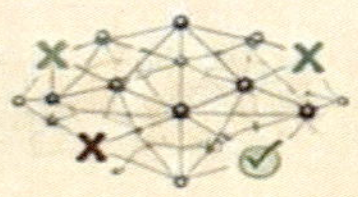

2015–2016
Precision and Data

Better cosmological and astrophysical data inform theoretical models.

2017–2018
New Dualities and Structures

Extended dualities and mathematical tools reveal deeper connections.

2019–2020
Quantum Information and Gravity

Insights from quantum information theory bridge gaps between fields.

2021–2023
Toward a New Synthesis

Unifying ideas grow stronger—though many mysteries remain.

2024 and Beyond
The Adventure Continues

New generations, new ideas, and new questions will guide us forward.

"The important thing is not to stop questioning. Curiosity has its own reason for existence." — Albert Einstein

THE HISTORY OF LOOP QUANTUM GRAVITY

1. CLASSICAL ROOTS

The Road to Quantum Gravity

Loop Quantum Gravity did not appear overnight. It grew from centuries of curiosity, bold ideas, and the relentless pursuit of understanding the deepest laws of nature.

From Einstein's vision of spacetime to the first hints that gravity must be quantum, a long road was traveled before the loop approach emerged.

> *"The important thing is not to stop questioning. Curiosity has its own reason for existence."*
> — Albert Einstein

MILESTONES THAT PAVED THE WAY

- **1687** Newton's Law of Universal Gravitation
- **1915** Einstein's General Relativity Unifies gravity and spacetime
- **1930s–1950s** Quantum Mechanics Transforms our view of the micro-world
- **1950s–1970s** The Search for Quantum Gravity Many paths explored, no clear answer
- **1980s** New mathematical tools and physical insights open new possibilities

FOUNDATIONS OF A NEW UNDERSTANDING

1. GENERAL RELATIVITY
1915

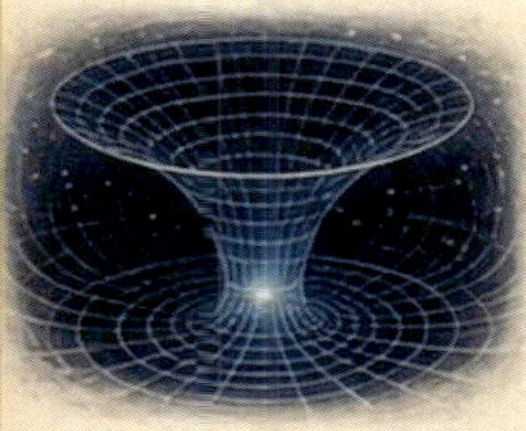

Space and time are not fixed arenas, but dynamic and intertwined. Matter tells spacetime how to curve, and spacetime tells matter how to move.

2. QUANTUM MECHANICS
1920s–1930s

At the smallest scales, reality is probabilistic and quantized. Observables come in discrete chunks, and uncertainty is fundamental.

3. THE CONFLICT
Mid 20th Century

General relativity governs the very large. Quantum mechanics governs the very small. Where they meet—our theories clash.

4. MANY APPROACHES
1950s–1980s

Many bold ideas emerged, each with its own promises and challenges. Among them, a path would arise that does not require extra dimensions.

5. A NEW POSSIBILITY
1980s–1990s

Inspired by geometry, quantum theory, and a desire for simplicity, a new framework began to take shape: Loop Quantum Gravity.

KEY INSIGHTS CARRIED FORWARD

Spacetime is relational, not absolute.

Geometry may be discrete at the most fundamental level.

New mathematics can reveal new physics.

Nature may be simpler—and more profound—than we imagine.

The loop approach seeks a background-independent and nonperturbative theory.

THE JOURNEY BEGINS

With these ideas and tools, a new path opens—one that seeks to quantize space itself, not on top of spacetime, but from within it.

This is where Loop Quantum Gravity begins.

THE STORY CONTINUES.

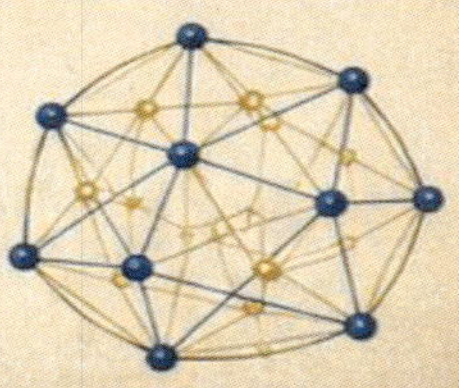

"Logic will get you from A to B. Imagination will take you everywhere." — Albert Einstein

Loop Quantum Gravity

2. The Need for Quantization

Gravity Meets Quantum Mechanics

General Relativity describes gravity as the curvature of spacetime on very large scales. Quantum Mechanics describes the behavior of matter and energy on the smallest scales.

But when we try to combine the two, we encounter contradictions—infinities, paradoxes, and a lack of predictive power.

To understand the universe completely, we need a quantum theory of gravity.

> *"A theory is all the more impressive the greater the simplicity of its premises, the more different kinds of things it relates, and the more extended its area of applicability."*
>
> — Albert Einstein

Why Quantize Gravity?

1. Infinities in General Relativity

Singularities at the centers of black holes and at the Big Bang lead to infinite densities and breakdowns of the theory.

2. Planck Scale Mysteries

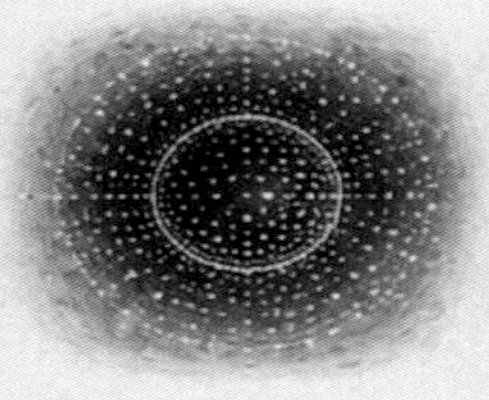

At the Planck scale ($\sim 10^{-35}$ meters), spacetime fluctuations make classical ideas of space and time meaningless.

3. Black Hole Information Paradox

Does information get destroyed in black holes? Quantum gravity must resolve this paradox.

4. The Early Universe

The first moments after the Big Bang require a theory that can unite quantum effects with dynamical spacetime.

5. A Complete Theory of Nature

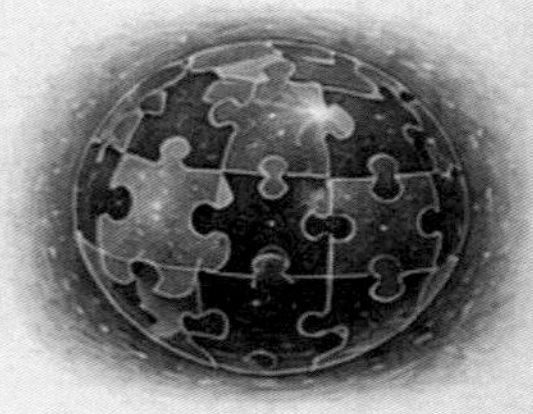

To understand everything—from galaxies to quarks—we need a single, consistent theory that includes gravity quantum mechanically.

Quantization: The Essential Idea

Just as energy comes in discrete quanta in quantum mechanics, spacetime itself must have a discrete structure.

In Loop Quantum Gravity, we quantize geometry, not fields on a fixed background. The fabric of spacetime becomes built from elementary building blocks.

Classical Geometry (Continuous) →

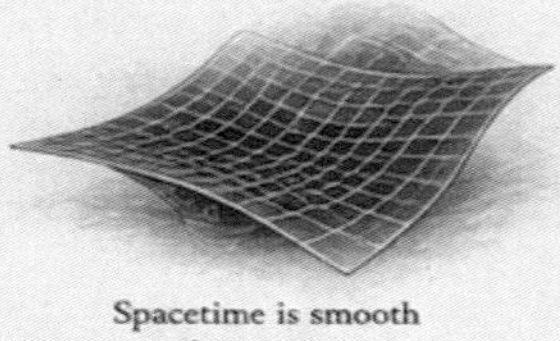

Spacetime is smooth and continuous.

Quantization →

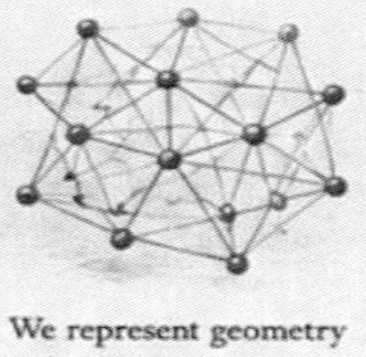

We represent geometry using connections and quantize them.

Discrete Geometry (Quantized) →

Areas and volumes have discrete spectra—no arbitrary smaller values.

Quantum Geometry

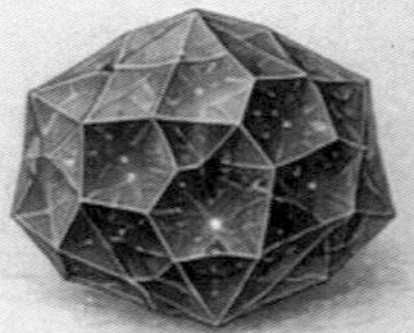

Spacetime is made of quanta of area and volume—a quantum geometry.

The Road Ahead

Quantizing gravity is one of the greatest intellectual challenges of our time. Loop Quantum Gravity offers a bold, background-independent path forward—rooted in mathematics, guided by principles, and driven by curiosity.

We are just beginning to glimpse the quantum nature of the universe. The journey has only just begun.

Looking Forward

- ★ Deepen our understanding of quantum geometry.
- ★ Connect LQG with cosmology and black holes.
- ★ Test predictions through observations and new technologies.
- ★ Inspire future generations of explorers and dreamers.

The History of Loop Quantum Gravity

3. Early Attempts (1930s–1960s)

Canonical Quantization Begins

To build a quantum theory of gravity, physicists first tried to quantize Einstein's General Relativity using the tools of quantum mechanics.

The challenge was immense: General Relativity is a theory of curved spacetime, not a field on a fixed background.

These early attempts laid the foundations for everything that followed.

Two great theories. One quantum future.

> *"We cannot expect to solve problems by using the same kind of thinking we used when we created them."*
>
> — Albert Einstein

Key Milestones and Ideas

1. Wheeler's Geometrodynamics
1930s

John Wheeler proposed that spacetime geometry is dynamic, not fixed. This insight inspired later efforts to quantize geometry itself.

2. Canonical Formulation
1950s

Physicists reformulated General Relativity in Hamiltonian form—the first step toward quantization.

3. ADM Formalism
1959–1962

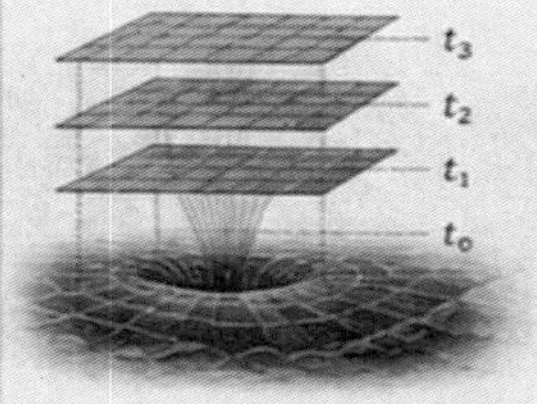

Arnowitt, Deser, and Misner split spacetime into space + time, enabling a Hamiltonian approach and defining constraints of the theory.

4. Constraints Emerge
1960s

The Hamiltonian and diffeomorphism constraints reflect the symmetry of General Relativity—key to any quantum theory of gravity.

5. No Clear Path
1960s

Canonical quantization proved extremely difficult. No consistent quantum theory of gravity had yet been achieved.

Lessons Learned

- ★ Spacetime is dynamic and must be quantized.
- ★ Symmetry and constraints are central to the theory.
- ★ Standard quantum field theory methods are not enough.
- ★ New mathematical and conceptual tools are required.
- ★ The path forward will require fresh ideas.

The questions were clear.
The answers would take decades.

Challenges Remaining

- ★ How to define quantum states of geometry?
- ★ How to implement the constraints without contradictions?
- ★ How to recover classical spacetime at large scales?
- ★ How to make contact with other quantum theories?
- ★ How to find a unique and predictive theory?

Preparing the Path for LQG

These early efforts, though incomplete, provided the language—constraints, canonical variables, and the Hamiltonian framework—that would later be essential to Loop Quantum Gravity.

They showed what must be done—and what must be avoided—as the search for quantum gravity continued.

> *"The most incomprehensible thing about the universe is that it is comprehensible."*
>
> — Albert Einstein

Every great theory begins with a question that refuses to disappear.

THE HISTORY OF LOOP QUANTUM GRAVITY

4. ASHTEKAR'S NEW VARIABLES (1980s)

A Breakthrough

In the early 1980s, Abhay Ashtekar introduced a revolutionary new way to describe General Relativity.

By using variables inspired by gauge theory and complex geometry, he revealed a hidden simplicity in Einstein's equations—one that would later become the foundation of Loop Quantum Gravity.

This was not a small step—it was a conceptual leap that changed everything.

New variables revealed a new path.

> *"The great triumph of physics is that its basic laws are simple, and yet they lead to a rich variety of phenomena."*
>
> — Abhay Ashtekar

THE REVOLUTIONARY IDEA

Ashtekar recast General Relativity using new variables:

1. SU(2) CONNECTION A_a^i

Instead of the metric, describe gravity using a connection—similar to the gauge fields of particle physics.

2. DENSITIZED TRIAD E_i^a

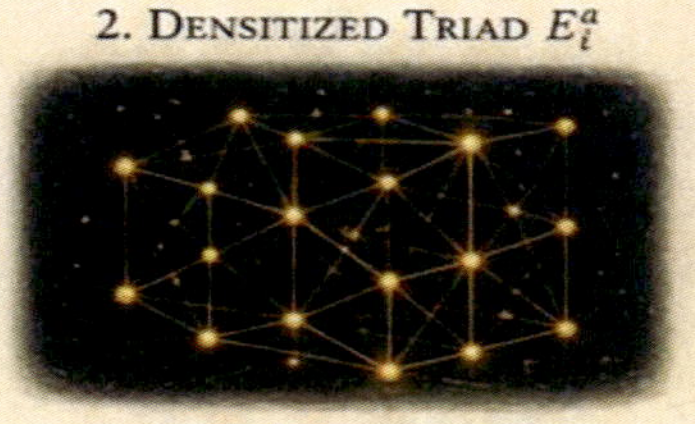

Instead of the metric's components, use a triad—an object that encodes the geometry of space.

3. CONSTRAINT EQUATIONS

- Gauss Constraint: $D_a E_i^a = 8\pi G\, \rho_i$
- Diffeomorphism Constraint:
  $F_{ab}^i E_i^b - A_a^i D_b E_i^b = 0$
- Hamiltonian Constraint:
  $\epsilon_i^{ijk} E_i^a E_j^b F_{ab}^k = 16\pi G\, \mathcal{H}$

Einstein's equations become elegant constraint equations, reminiscent of gauge theories.

KEY BREAKTHROUGHS

1. COMPLEX VARIABLES

Ashtekar used complex variables, making the equations simpler and more symmetric.

2. GAUGE THEORY STRUCTURE

Gravity now looked like a gauge theory—opening the door to powerful tools.

3. NON-PERTURBATIVE POTENTIAL

The new variables allowed background-independent, non-perturbative approaches.

4. CANONICAL FORMULATION

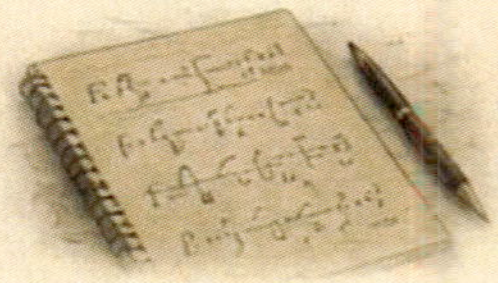

Ashtekar developed a full Hamiltonian (canonical) formulation of gravity.

5. QUANTIZATION BECOMES VIABLE

With the new variables, quantization became a tangible possibility.

A CHANGE OF LANGUAGE

Just as Newton's laws replaced ancient ideas, Ashtekar's variables replaced the old language of gravity.

A new language reveals what was always there.

THE IMPACT

Ashtekar's insight provided the mathematical groundwork for Loop Quantum Gravity. It showed that space, geometry, and gravity could be described using variables that are natural, simple, and deeply connected to the quantum world.

LOOKING FORWARD

- ★ The new variables united gravity with modern physics.
- ★ They inspired a new generation of researchers.
- ★ They laid the foundation for Loop Quantum Gravity.
- ★ They showed that beauty and simplicity often lead to truth.

From insight comes possibility. From possibility, discovery.

> *"The important thing is not to stop questioning. Curiosity has its own reason for existence."*
> — Albert Einstein

5. CONNECTION VARIABLES
REFORMULATE GRAVITY

A New Language for Space, Time, and Curvature

In the late 1980s, researchers discovered that General Relativity could be rewritten using connection variables—mathematical objects that behave like gauge fields.

This reformulation revealed a deep connection between gravity and the other fundamental forces of nature.

It also opened the door to using the powerful tools of gauge theory and quantum field theory.

This new language would become the backbone of Loop Quantum Gravity.

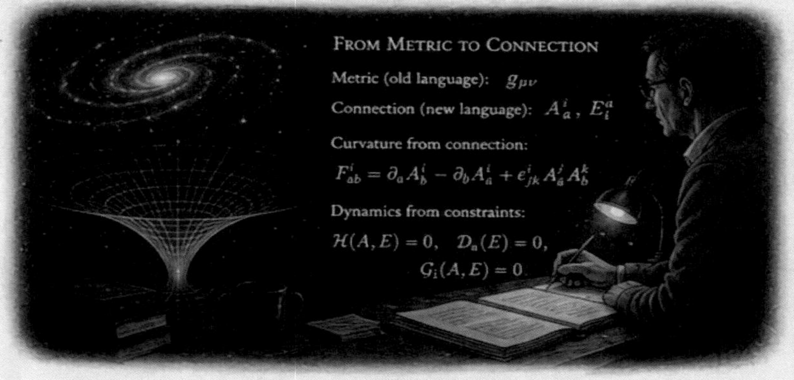

"The universe speaks to us in many languages. We must find the one that reveals its deepest symmetry."

— Carlo Rovelli

WHY CONNECTION VARIABLES?

1. GAUGE STRUCTURE

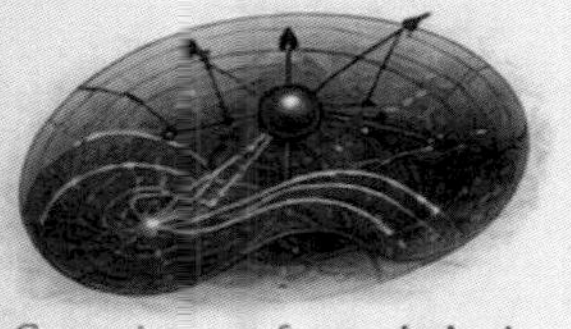

Connections transform under local gauge transformations, just like the electromagnetic and weak forces. Gravity is revealed as a gauge theory of the Lorentz group.

2. SEPARATES GEOMETRY FROM DYNAMICS

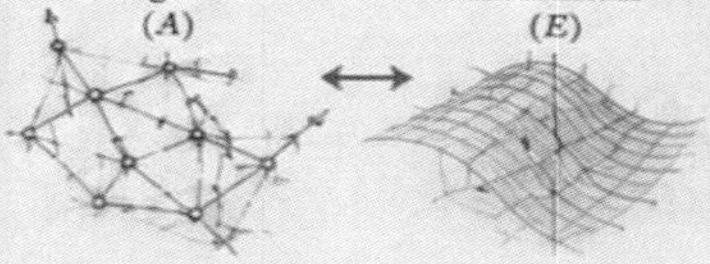

The connection A encodes how space is "twisted," while the densitized triad E encodes "how much." Together they capture the geometry of space.

3. PREPARES FOR QUANTIZATION

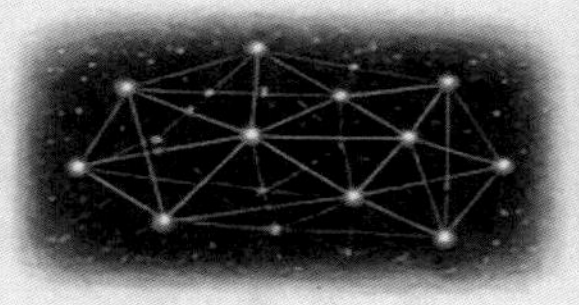

The variables (A, E) lead naturally to well-defined Poisson brackets and a path to quantization using holonomies and fluxes.

4. UNIFIES NATURE

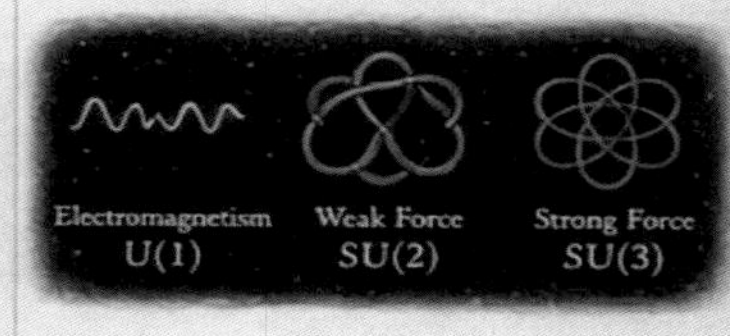

Using connection variables shows gravity shares the same mathematical structure as the other fundamental interactions.

THE MATHEMATICAL CORE

BASIC VARIABLES

Connection: A^i_a
(SU(2) connection)

Densitized triad: E^a_i
(vector density of weight +1)

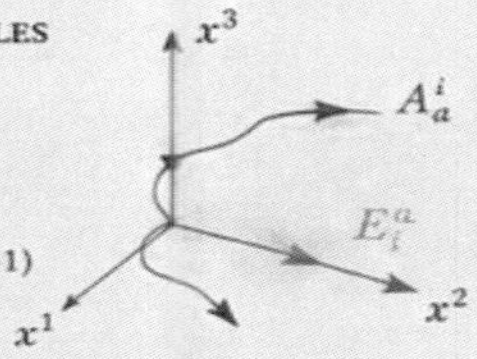

POISSON BRACKET

$$\{A^i_a(x), E^b_j(y)\} = 8\pi G\gamma\, \delta^i_j \delta^b_a \delta^{(3)}(x,y)$$

γ is the Immirzi parameter, a dimensionless constant.

CONSTRAINTS OF GENERAL RELATIVITY

Gauss (internal rotations):
$$G_i = \mathcal{D}_a E^a_i = 0$$
Diffeomorphism (spatial):
$$V_a = E^b_i F^i_{ab} - (1+\gamma^2) K_a = 0$$
Hamiltonian (time evolution):
$$H = \frac{\epsilon^{ijk}_j E^a_i E^b_j}{\sqrt{\det E}} F^k_{ab} - (1+\gamma^2) E^a_i E^b_j K^i_a K^j_b = 0$$

FROM VARIABLES TO GEOMETRY

1. HOLONOMIES
Paths carry information about the connection.

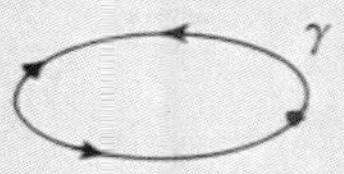

$$h_\gamma[A] = \mathcal{P}\exp\left(\int_\gamma A\right)$$

2. FLUXES
Surfaces measure the triad field.

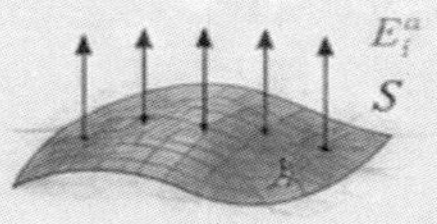

$$E_i(S) = \int_S E^a_i n_a\, d^2\sigma$$

3. DISCRETE SPECTRUM
Areas and volumes have quantized values.

$$A = 8\pi\gamma\,\ell_P^2 \sqrt{j(j+1)}$$
$$j = 0, \frac{1}{2}, 1, \frac{3}{2}, \dots$$

4. QUANTUM GEOMETRY
Space itself is discrete at the smallest scale.

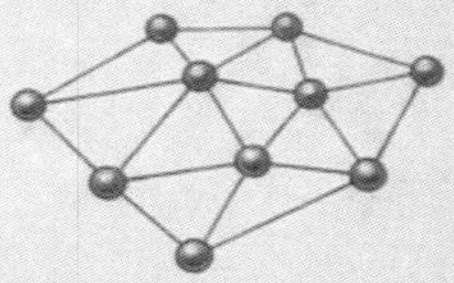

WHAT THIS ACHIEVEMENT MADE POSSIBLE

- ★ A background-independent formulation.
- ★ A non-perturbative approach to quantum gravity.
- ★ A bridge between General Relativity and quantum theory.
- ★ The foundation for the Loop Quantum Gravity program.

THE JOURNEY CONTINUES

- ★ Build the quantum theory.
- ★ Explore black holes and the Big Bang.
- ★ Uncover the nature of spacetime.
- ★ Seek the single, beautiful theory that unites all things.

"The greatest step forward is not in solving everything, but in finding the right way to ask." — Carlo Rovelli

6. Birth of Loop Quantum Gravity

Late 1980s

In the late 1980s, Carlo Rovelli and Lee Smolin proposed a radically new, non-perturbative approach to quantum gravity.

Instead of quantizing spacetime itself, they quantized the geometry of space using connections and loops—leading to what is now known as Loop Quantum Gravity.

It was a bold idea, born from insight, subtle mathematics, and a deep belief that space is more fundamental than spacetime.

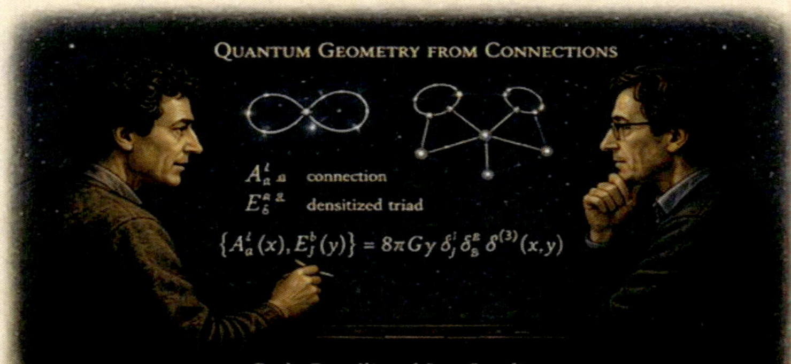

Carlo Rovelli and Lee Smolin
Proposing a new path forward.

> *"We knew we were building something different.*
> *Not a small correction, but a new framework.*
> *A new way to see the quantum world."*
>
> — Carlo Rovelli
> (Years later)

The Revolutionary Idea

1. Quantize the Geometry of Space

Space is made of discrete building blocks. Geometry itself becomes quantum.

2. Use Connections and Loops

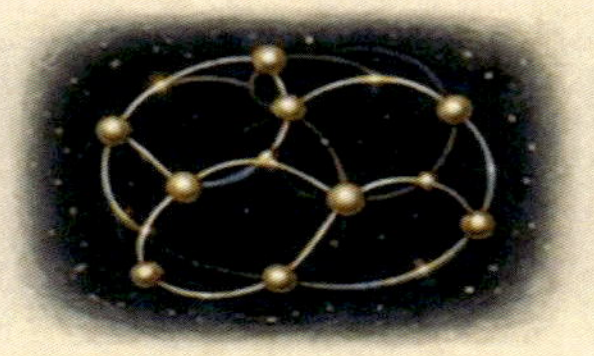

Connections describe how geometry changes. Loops capture the essence of curvature and holonomy.

3. New Mathematical Tools

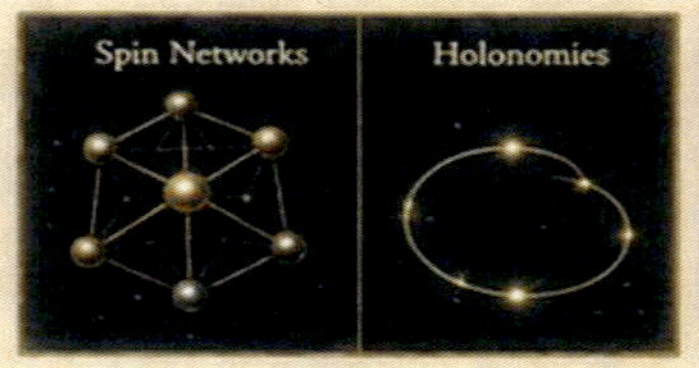

Spin networks describe quantum states of space. Holonomies encode parallel transport around loops.

4. Background Independence

No fixed spacetime background. Geometry emerges from the quantum relationships between fundamental degrees of freedom.

Key Features of Loop Quantum Gravity

Discrete Spectrums of Geometry

Areas and volumes have discrete spectra.
Space is granular.

Non-Perturbative and Background Independent

No fixed background.
The theory works non-perturbatively.

New Kind of Quantum States

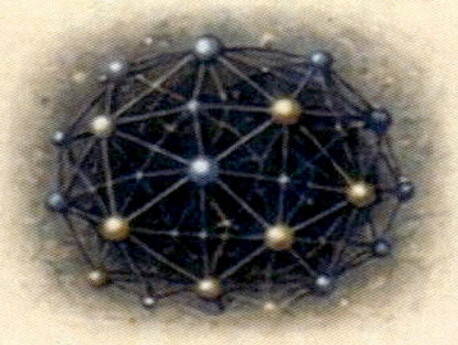

Spin networks provide a basis for the quantum geometry of space.

Rich Mathematical Structure

Uses tools from differential geometry, topology, and operator algebras.

Potential to Resolve Singularities

Quantum geometry may resolve the Big Bang and black hole singularities.

A New Vision of the Universe

Loop Quantum Gravity suggests that at the deepest level, space is woven from quantum threads. The smooth spacetime of General Relativity emerges from this discrete quantum fabric.

A new picture of reality is beginning to emerge.

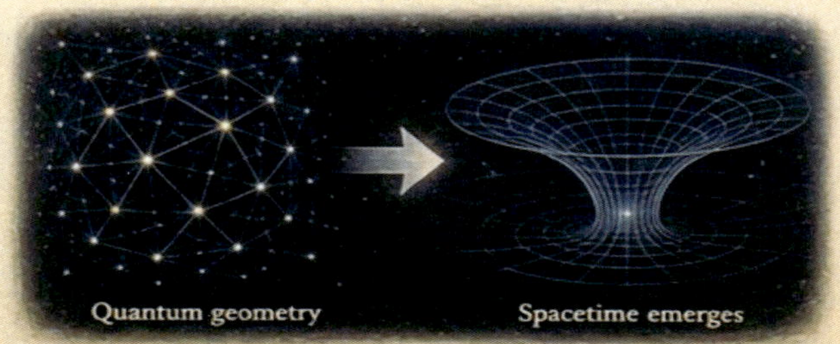

What LQG Promises

- ★ A consistent theory of quantum gravity
- ★ Insights into the origin of spacetime
- ★ A resolution of black holes and the Big Bang
- ★ A new roadmap for fundamental physics
- ★ A deeper understanding of reality itself

The birth of Loop Quantum Gravity was not the end of the story—it was the beginning.

"Logic will get you from A to B. Imagination will take you everywhere." — Albert Einstein

7. SPIN NETWORKS

The Quantum Geometry

Once geometry is quantized, space itself becomes built from discrete, combinatorial structures called spin networks.

They capture the quantum states of geometry—telling us how space is woven.

Spin networks are the basis states of Loop Quantum Gravity. They changed how we think about space forever.

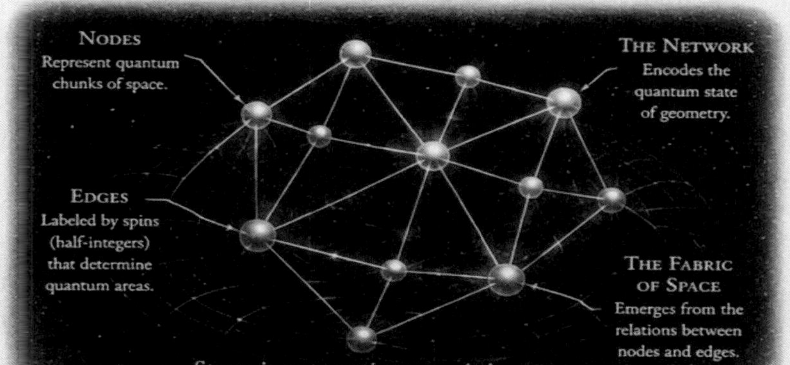

Space is not continuous—it is woven.

> *"A spin network is a quantum state of the gravitational field. It describes quantum space."*
>
> — Abhay Ashtekar

WHAT IS A SPIN NETWORK?

1. GRAPH STRUCTURE

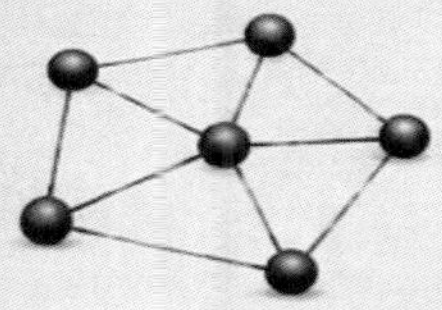

A spin network is a graph with nodes (vertices) and edges (links). It has no background—only relations.

2. SPIN LABELS

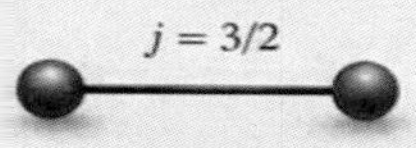

Each edge is labeled by a spin $j = 0, 1/2, 1, 3/2, \ldots$ These spins determine quantized areas.

3. INTERTWINERS AT NODES

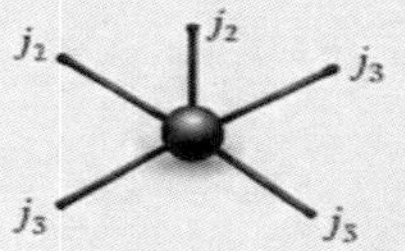

At each node, spins meet. Intertwiners ensure the quantum states are gauge-invariant.

4. QUANTIZED AREAS

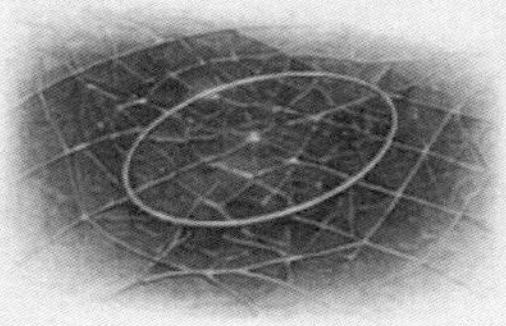

The area of a surface pierced by an edge is quantized:

$$A = 8\pi\gamma\, \ell_P^2 \sqrt{j(j+1)}$$

5. QUANTIZED VOLUMES

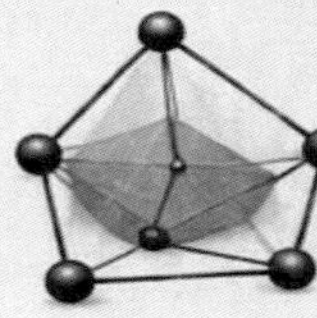

Volumes are quantized at nodes, depending on the incident spins and intertwiners.

WHY SPIN NETWORKS MATTER

STATES OF GEOMETRY

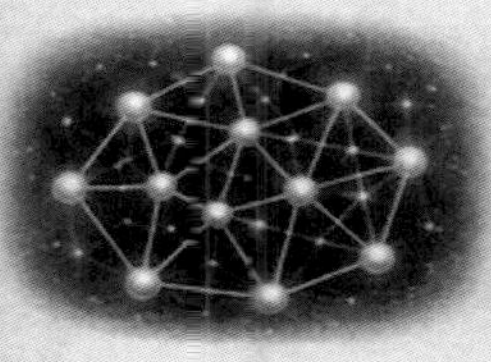

Spin networks provide a basis for the quantum states of spatial geometry.

BACKGROUND INDEPENDENT

They exist without any fixed background spacetime. Only relations matter.

DYNAMICAL

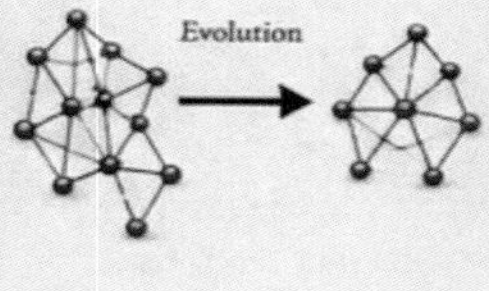

Spin networks evolve via dynamical rules, describing quantum geometry in time.

FOUNDATION FOR QUANTUM SPACETIME

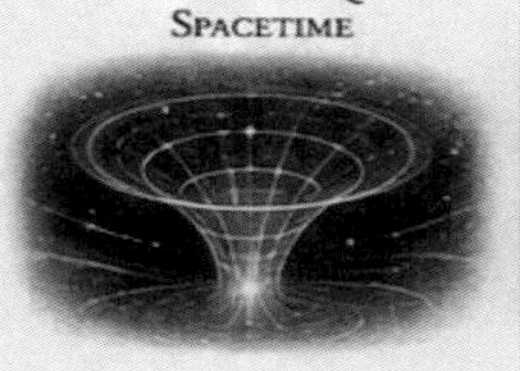

They are the building blocks from which spacetime—and gravity—emerge.

LINK TO OBSERVABLES

$$\text{Area} = 8\pi\gamma\, \ell_P^2 \sqrt{j(j+1)}$$

$$\text{Volume} \sim \gamma^{3/2}\, \ell_P^2 \sqrt{V(j_i, I)}$$

They give rise to quantized geometric observables, with testable signatures.

A NEW PICTURE OF SPACE

Classical geometry is replaced by a quantum fabric woven from discrete threads of relation.

Space is not made of points. It is made of connections.

Spin networks show us that geometry itself is quantum.

This is the quantum nature of space.

FROM CONTINUOUS
THE SMOOTH FABRIC

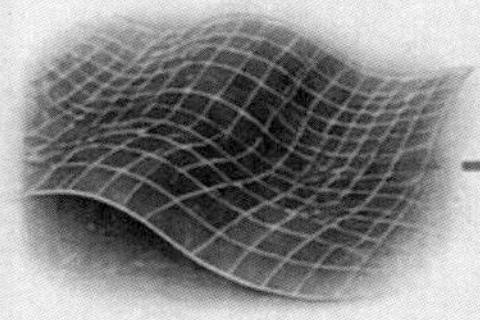

TO DISCRETE
THE WEAVE

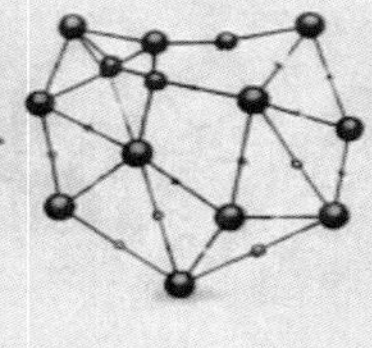

TO DYNAMICAL
THE LIVING NETWORK

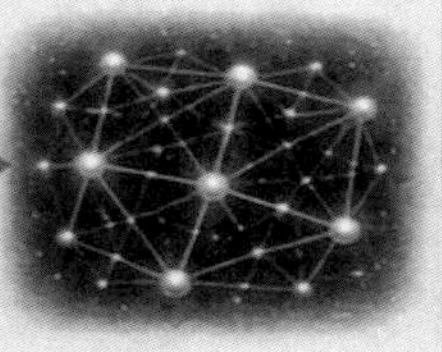

From smooth space, to quantum weave, to a living geometry.

LOOKING AHEAD

- ★ Spin networks are the language of LQG.
- ★ They open the door to the quantum dynamics of space.
- ★ They guide the search for the quantum nature of spacetime and gravity.
- ★ They connect geometry, physics, and reality in a new way.

"Reality is relational. Space is woven. Geometry is quantum." — Abhay Ashtekar

8. Background Independence

A Key Principle

General Relativity is background independent—there is no fixed spacetime stage. Geometry dynamically arises from matter and energy.

Loop Quantum Gravity takes this seriously at the quantum level. It does not quantize fields on a background. It quantizes the geometry itself.

This principle lies at the heart of the theory.

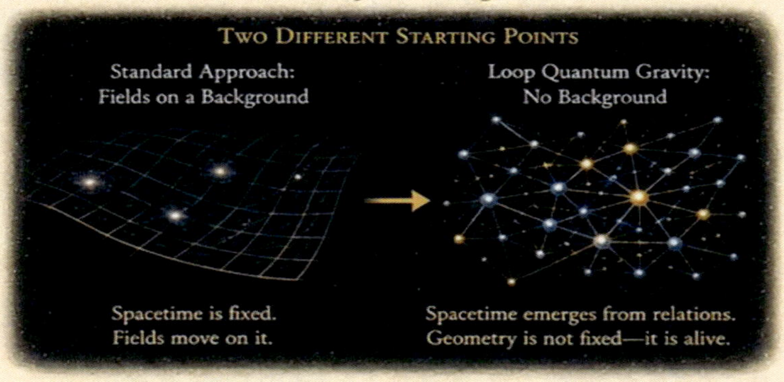

"Background independence is not a technical detail. It is a statement about what reality is made of."

— Carlo Rovelli

What Does Background Independence Mean?

1. No Fixed Spacetime

There is no fixed metric waiting in the background. Spacetime is not a container.

2. Relational Reality

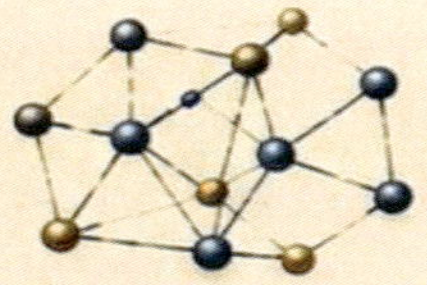

Only relations are fundamental. Geometry arises from them.

3. Diffeomorphism Invariance

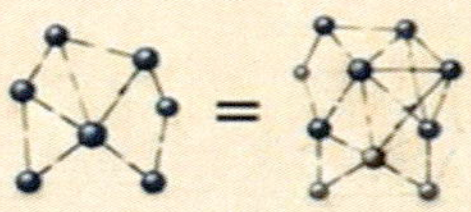

Physical predictions do not depend on arbitrary coordinate choices. Only the relational structure matters.

4. Dynamical Geometry

Geometry changes, fluctuates, and evolves. There is no background to fluctuate on.

5. Universal Principle

Background independence unifies gravity with quantum principles in a consistent way.

How LQG Realizes Background Independence

1. Connections Not Metrics

LQG uses connections (A) instead of the metric. Connections encode geometry without referring to a background.

2. Holonomies

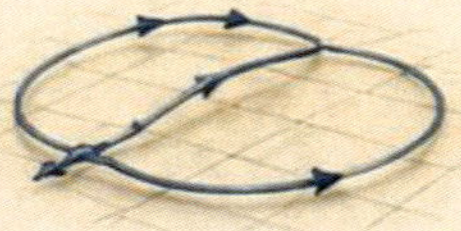

Holonomies describe how vectors rotate when transported along loops. They are background independent.

3. Densitized Triads

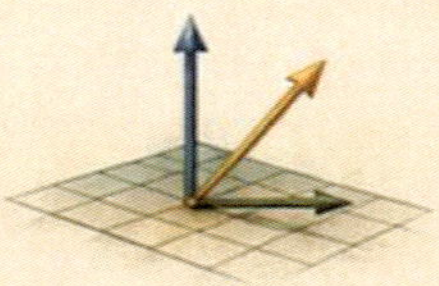

Triads encode spatial geometry (densitized). They are fields on space, not on spacetime.

4. Constraints Generate Gauge

$$
\begin{aligned}
G_i &= D_a E^a = 0 \\
V_a &= E_i^b F_{ab}^i = 0 \\
H &= \epsilon^{ijk} E_i^a E_j^b F_{ab}^k \\
&\quad - (1+\gamma^2)\, K_{[a}^i E_{b]}^j E_j^k = 0
\end{aligned}
$$

The Gauss, Diffeomorphism, and Hamiltonian constraints ensure background independence at the quantum level.

5. Spin Network States

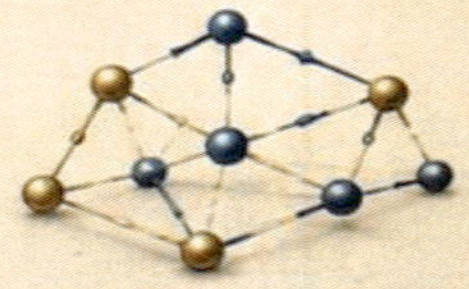

The quantum states of geometry are spin networks—combinatorial, background-independent structures.

Why It Matters

- ★ It is essential for a true theory of quantum gravity.
- ★ It avoids reliance on unphysical background structures.
- ★ It aligns with the spirit of General Relativity.
- ★ It may help resolve deep puzzles like the problem of time.
- ★ It points toward a deeper understanding of spacetime itself.

"We do not quantize spacetime. We let spacetime be born from quantum relations."

— Carlo Rovelli

Looking Ahead

Background independence is more than a principle—it is a compass. It guides every step of Loop Quantum Gravity toward a deeper, more fundamental description of reality, where space and time are not things, but patterns of connection.

The Journey Continues

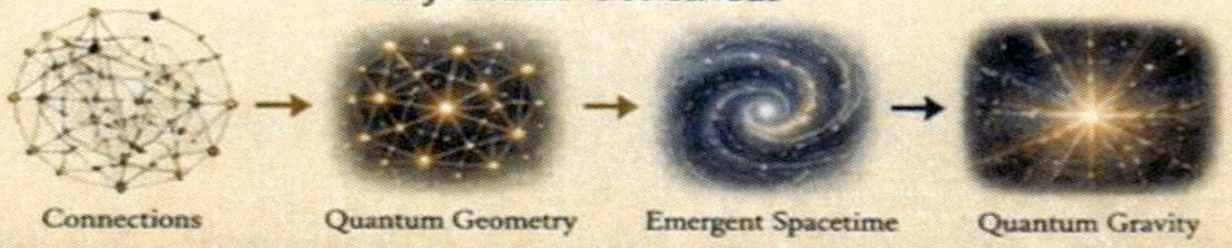

Key Insight

When we let go of the background, we discover that the universe is not built on a stage— it is a dance of relations.

Relativity becomes quantum.

"The aim of science is not to open the door to infinite wisdom, but to set a limit to infinite error." — Bertolt Brecht

9. QUANTIZED AREAS AND VOLUMES

Space Has a Discrete Structure

In the 1990s, Loop Quantum Gravity revealed something astonishing: space itself is quantized.

Areas and volumes do not vary smoothly. They come in discrete "chunks," much like the energy levels of an atom.

This result emerged from the mathematics of spin networks and the operators of quantum geometry.

QUANTUM GEOMETRY REVEALS DISCRETE SPECTRA

AREAS ARE QUANTIZED

Eigenvalues:

$$A_j = 8\pi\gamma\, \ell_P^2 \sqrt{j(j+1)}$$

Each edge labeled by spin j contributes a discrete amount of area when it pierces a surface.

VOLUMES ARE QUANTIZED

Eigenvalues:

$$V_n = \ell_P^3 \sqrt{|n|}\, f(n)$$

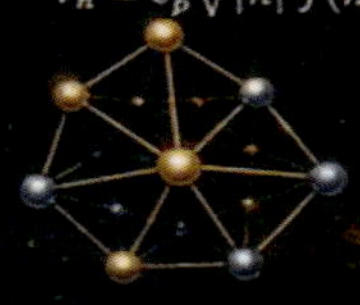

Each node (intertwiner) contributes a discrete amount of volume to the region it defines.

> *"The spectra of area and volume operators are discrete. Space is not a continuum but a quantum geometry."*
>
> — Abhay Ashtekar

HOW QUANTIZATION ARISES

1. CLASSICAL SURFACE

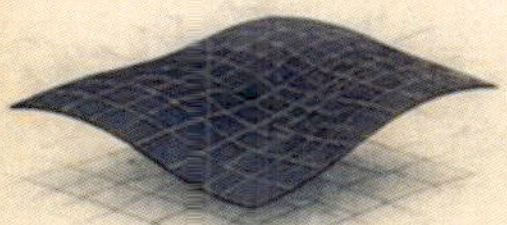

A smooth surface cuts through space. Its area can vary continuously.

2. PUNCTURES BY SPIN NETWORK EDGES

Edges of a spin network puncture the surface at points. Each carries a spin j.

3. QUANTUM CONTRIBUTIONS ADD UP

$$A = \sum_i 8\pi\gamma\, \ell_P^2 \sqrt{j_i(j_i+1)}$$

Each puncture contributes a discrete amount of area. The total area is quantized.

4. DISCRETE AREA EIGENVALUES

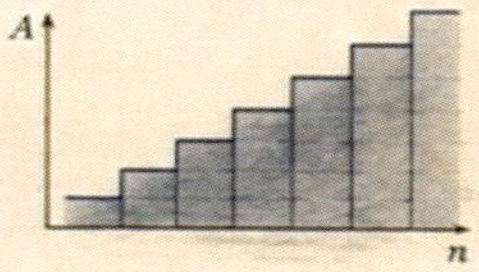

Areas come in a discrete spectrum. There is a smallest non-zero area.

5. DISCRETE VOLUME FROM NODES

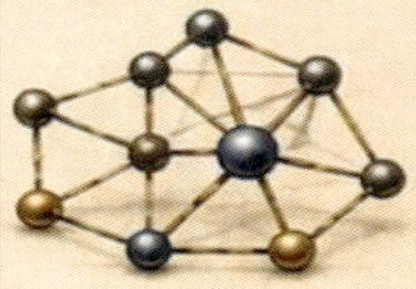

Nodes join edges (intertwiners). Their quantum states give rise to discrete volumes.

KEY RESULTS IN LQG

1. AREA QUANTIZATION

$$A_j = 8\pi\gamma\, \ell_P^2 \sqrt{j(j+1)}$$

Area eigenvalues depend on the Barbero–Immirzi parameter γ and the spin $j = 1/2, 1, 3/2, \ldots$

Smallest non-zero area:

$$A_{\min} = 4\pi\gamma\, \ell_P^2 \sqrt{3}$$

2. VOLUME QUANTIZATION

$$V_n = \ell_P^3 \sqrt{|n|}\, f(n)$$

Volume eigenvalues depend on the intertwiners at nodes and their combinatorial structure.

Smallest non-zero volume is of order $\gamma^{3/2}\, \ell_P^3$.

3. IMPORTANCE OF γ

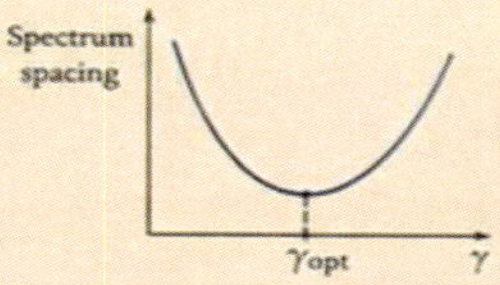

The Barbero–Immirzi parameter sets the scale of quantum geometry. Its value is fixed by black hole entropy.

4. BACKGROUND INDEPENDENT

Quantization does not rely on any background metric. The discreteness arises from the quantum states themselves.

5. PLANCK-SCALE STRUCTURE

At the Planck scale, space is a granular, foamy network of quantized building blocks.

WHAT THIS MEANS

CONTINUOUS VIEW (CLASSICAL)

In classical physics, space and time are smooth and continuous.

→ **QUANTUM VIEW (LQG)**

In Loop Quantum Gravity, space is built from discrete quantum building blocks.

→ **AT THE PLANCK SCALE**

Zooming in reveals a fundamental discreteness. There is no smaller structure.

→ **THIS LEADS TO...**

New insights into black holes, the Big Bang, and the nature of spacetime itself.

A NEW UNDERSTANDING

"The fabric of space is woven from threads of quantum geometry. Reality is granular at its most basic level."

LOOKING AHEAD

Quantized areas and volumes are just the beginning. They provide the foundation for understanding evolution, dynamics, and the emergence of spacetime.

THE JOURNEY CONTINUES

- ★ Quantized geometry
- ★ Dynamical evolution
- ★ Black hole physics
- ★ Cosmology and the Big Bang
- ★ Toward a complete theory

The path of discovery is far from over.

"The most incomprehensible thing about the universe is that it is comprehensible." — Albert Einstein

10. THE GAUSS CONSTRAINT

Ensuring Gauge Invariance

Gauge invariance is a fundamental symmetry of nature. In Loop Quantum Gravity, it is enforced by the Gauss constraint, which guarantees that our description of space is independent of local "gauge rotations" at each point.

It ensures that only physical, gauge-invariant states are allowed—eliminating unphysical degrees of freedom.

The Gauss constraint is the quantum statement of charge conservation and local symmetry.

GAUGE TRANSFORMATIONS AT A NODE

A gauge transformation rotates all internal directions at a node without changing physical content.

> *"The Gauss constraint ensures that physics does not depend on our arbitrary choices of internal frames. It is the guardian of local symmetry."*
>
> — Carlo Rovelli

THE GAUSS CONSTRAINT IN LQG

1. Internal Rotations

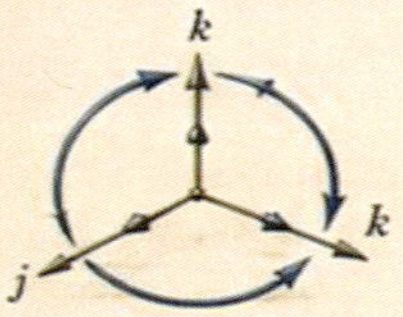

At each point, we can rotate internal directions (SU(2)) without changing physics.

2. Flux Operators

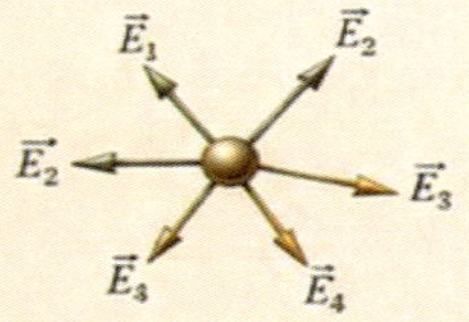

Edges carry fluxes $\vec{E}_e$ of the triad field. They represent the "electric field" of geometry.

3. The Gauss Constraint

$$G^i = \sum_{e \text{ at } n} E^i_e = 0$$

At every node n, the sum of outgoing fluxes must vanish. This is the Gauss constraint.

⇒ No net "charge" at a node.

4. Generates Gauge Transformations

$$[G^i, A^j_a] = \epsilon^{ij}{}_k A^k_a \delta^{(3)}(x, x_n)$$

$$[G^i, E^a_j] = \epsilon^{ik}_j E^a_k \delta^{(3)}(x, x_n)$$

The constraint generates infinitesimal SU(2) rotations of the connection and triad at the node n.

5. Physical States

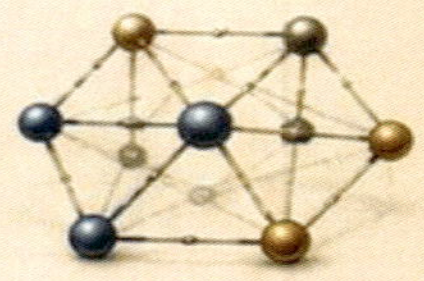

Physical (gauge-invariant) states are annihilated by the Gauss constraint: $G^i|\psi\rangle = 0$
They encode true geometry.

IMPLICATIONS AND IMPORTANCE

1. Gauge Invariance

Ensures results do not depend on arbitrary internal choices. Only relational information is physical.

2. Consistency

Together with the diffeomorphism and Hamiltonian constraints, it ensures a consistent, complete theory.

3. Geometric Meaning

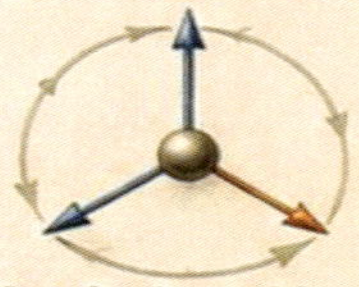

No preferred internal direction exists at a point—only the relations between edges have meaning.

4. Quantum Interlocking

Gauge invariance intertwines edges at a node. Intertwines are the quantum states that satisfy $G^i = 0$.

5. Foundation of LQG

The Gauss constraint is one of the pillars of Loop Quantum Gravity, ensuring symmetry, consistency, and physicality.

A SIMPLE EXAMPLE: THREE-VALENT NODE

Consider a node with three edges, with spins j_1, j_2, j_3.
Gauge invariance requires that their vector sum can be zero.
This is possible iff the spins can form a triangle:

$|j_1 - j_2| \leq j_3 \leq j_1 + j_2$
(and cyclic permutations)

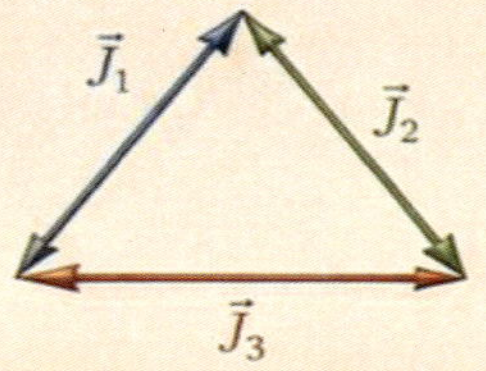

⇔

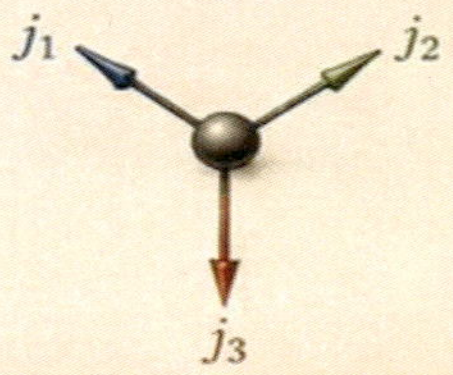

Intertwiners are the gauge-invariant states at the node. For three-valent nodes, there is a unique (up to phase) intertwiner when the triangle condition is satisfied.

LOOKING AHEAD

- ★ The Gauss constraint works hand in hand with the other constraints.
- ★ Together they shape the quantum dynamics of spacetime.
- ★ They guide us toward a universe that is consistent, symmetric, and deeply quantum.

THE JOURNEY CONTINUES

- ★ Quantize the Hamiltonian constraint (the dynamics).
- ★ Solve the constraints together.
- ★ Understand time, change, and evolution in quantum spacetime.
- ★ Move ever closer to a complete theory of quantum gravity.

"The highest form of understanding is to see the unity in diversity." — Einstein

11. The Hamiltonian Constraint

The Dynamics of Quantum Geometry

To describe dynamics, Loop Quantum Gravity promotes the Hamiltonian constraint—the generator of "time" evolution in General Relativity—to a quantum operator.

Physical states of the universe must be annihilated by all constraints, including the Hamiltonian constraint:

$$\hat{H}|\psi\rangle = 0$$

Solutions are called "spin foam histories."

Evolution Without Time

In LQG, time is not fundamental. Change is encoded in the relations between quantum states of geometry.

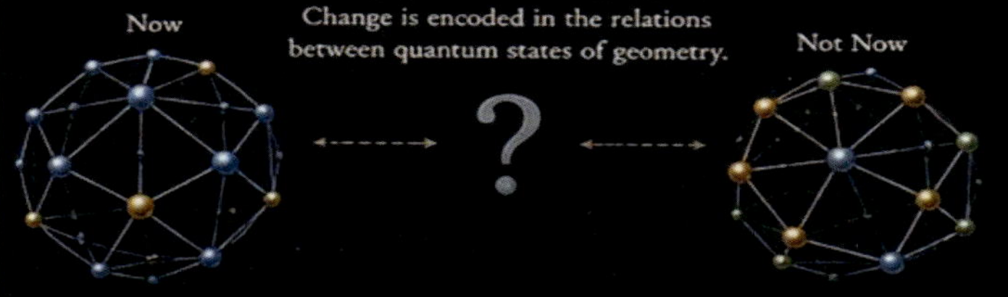

The Hamiltonian constraint links "now" to "not now" without reference to an external time parameter.

> *"The Hamiltonian constraint is the engine of quantum evolution, yet it does not rely on time as we know it. It generates change without time."*
>
> — Abhay Ashtekar

Building the Hamiltonian Constraint in LQG

1. Classical Origin

$H = H_E + H_L + H_M$

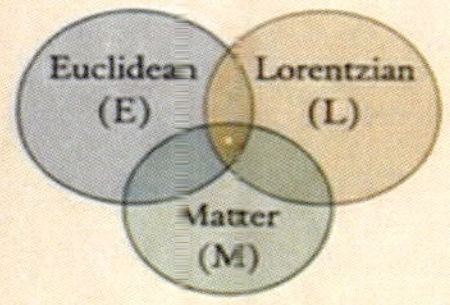

The classical Hamiltonian decomposes into Euclidean, Lorentzian, and matter parts.

2. Regularization

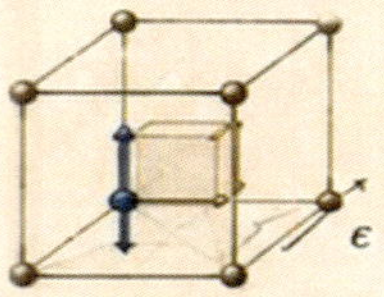

Space is discretized into small cells. Curvature is represented by holonomies around loops.

3. Quantization

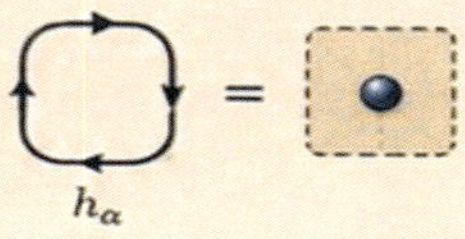

Holonomies and fluxes become operators. Curvature is encoded in holonomies around loops.

4. Constraint Operator

$\hat{H}(N) = \sum_v N(v)\hat{H}_v$

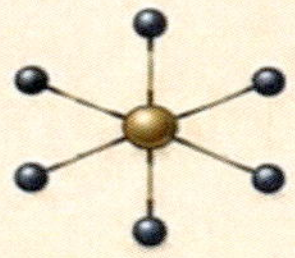

The Hamiltonian is a sum over nodes v, smeared with lapse function $N(v)$.

5. Annihilation Condition

$\hat{H}(N)|\psi\rangle = 0$

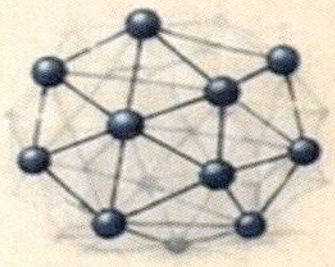

Physical states are those annihilated by the Hamiltonian for all lapse functions $N(v)$.

Action on Spin Networks

The Hamiltonian constraint acts at a node by creating, moving, or reshaping edges. It changes the spin network, generating new quantum geometries.

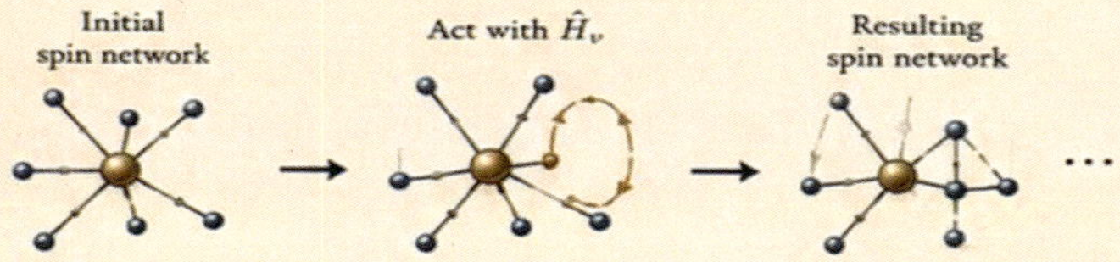

The operator can:

- ★ Create or annihilate loops
- ★ Change spins on edges
- ★ Change the intertwiner at the node

Key Properties

- ★ **Background independent**
  The Hamiltonian does not refer to any external spacetime.
- ★ **Diffeomorphism invariant**
  It treats all directions equally.
- ★ **Non-local in a small region**
  It acts only at nodes but can create closed loops.
- ★ **Highly non-trivial**
  Its precise definition and solutions remain one of the deepest challenges.

Spin Foams: Histories of Geometry

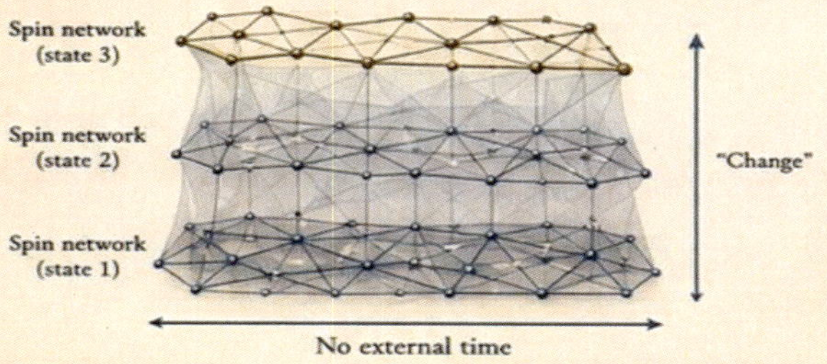

Spin foams represent quantum histories—successions of spin networks connected by local transitions generated by the Hamiltonian constraint.

They describe a universe in quantum becoming, without time.

Dynamics is not evolution in time. It is the web of quantum relations between possible geometries.

Looking Ahead

- ★ Solving the Hamiltonian constraint is the central challenge of LQG.
- ★ It will unlock the quantum dynamics of spacetime.
- ★ The path forward is mathematical, conceptual, and computational.
- ★ Persistence brings us closer to the quantum theory of everything.

The Journey Continues

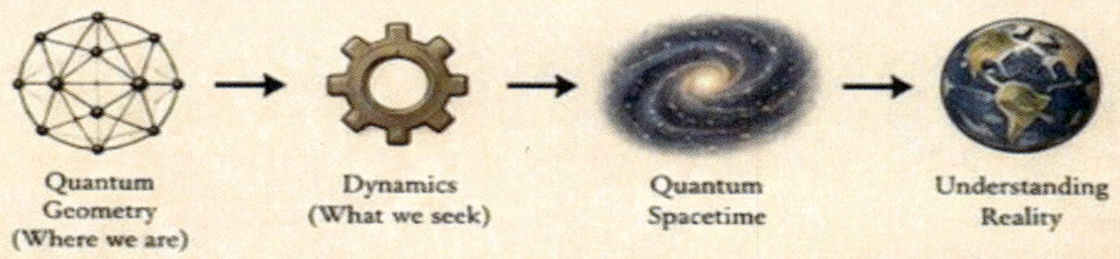

Step by step, constraint by constraint, we approach the truth.

"In the struggle with the Hamiltonian, we are learning how the universe becomes." — Abhay Ashtekar

12. SPIN FOAMS

Histories of Spin Networks

Spin networks describe quantum geometry at a single "moment." But the universe evolves.

Spin foams provide a way to represent that evolution—the history of spin networks in spacetime.

They are the path integral formulation of Loop Quantum Gravity.

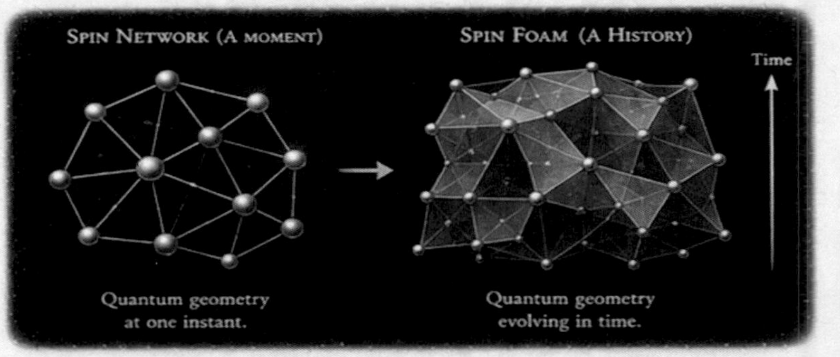

> *"A spin foam is a timeline of quantum geometries. It shows how space itself can emerge, evolve, and transform."*
>
> — Carlo Rovelli

WHAT IS A SPIN FOAM?

1. 2-COMPLEX

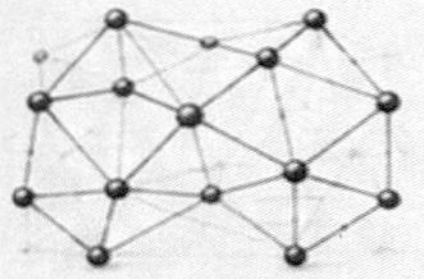

A spin foam is built on a 2-complex: vertices, edges, and faces.

2. LABELS

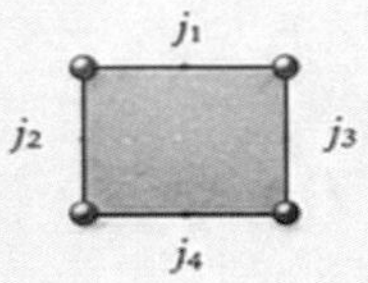

Faces are labeled by spins (*j*). Edges are labeled by intertwiners (*i*).

3. INTERTWINERS

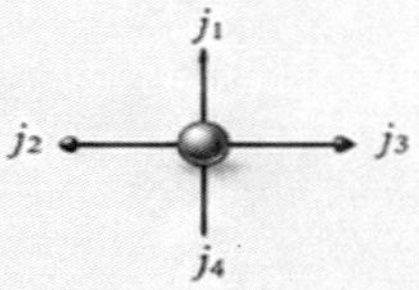

Intertwiners encode how quantum geometries are connected across time.

4. AMPLITUDES

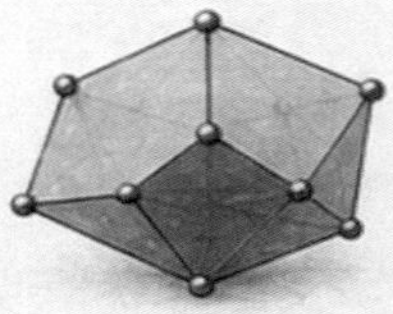

Each vertex (foam) contributes an amplitude. The full foam has a total amplitude.

5. SUM OVER HISTORIES

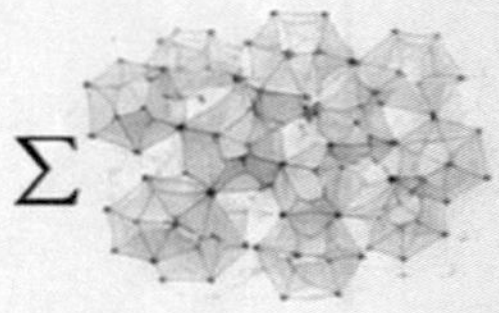

We sum over all possible spin foams—histories of quantum geometry.

HOW SPIN FOAMS DESCRIBE EVOLUTION

INITIAL SPIN NETWORK — SPIN FOAM (THE HISTORY) — FINAL SPIN NETWORK

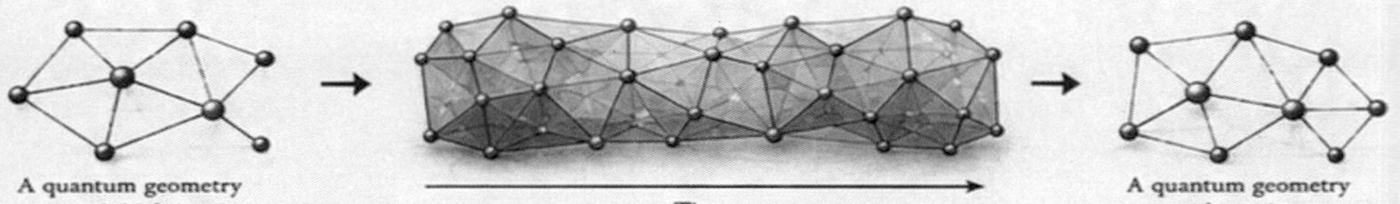

A quantum geometry at an initial time. — A quantum geometry at a later time.

Spin foams interpolate between spin networks, providing a covariant (diffeomorphism-invariant) way to describe quantum evolution.

KEY ASPECTS OF SPIN FOAMS

1. COVARIANT

No preferred time. Spin foams are fully background independent and Lorentz covariant.

2. LOCAL

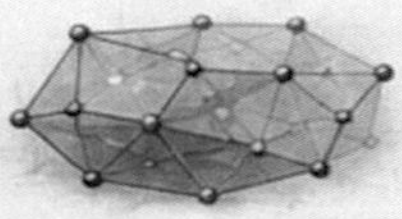

Amplitudes are local: determined by the spins and intertwiners around each vertex.

3. FINITENESS

For fixed boundary data, the sum over spin foams is expected to be finite.

4. RECOVERS GEOMETRY

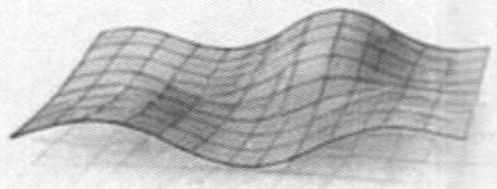

In suitable limits, smooth spacetime geometry emerges from the quantum foam.

5. PATH INTEGRAL

$$Z = \sum_{\text{foams}} A(\text{foam})$$

The transition amplitude is obtained by summing over all spin foam histories.

A SIMPLE EXAMPLE: CYLINDRICAL SPIN FOAM

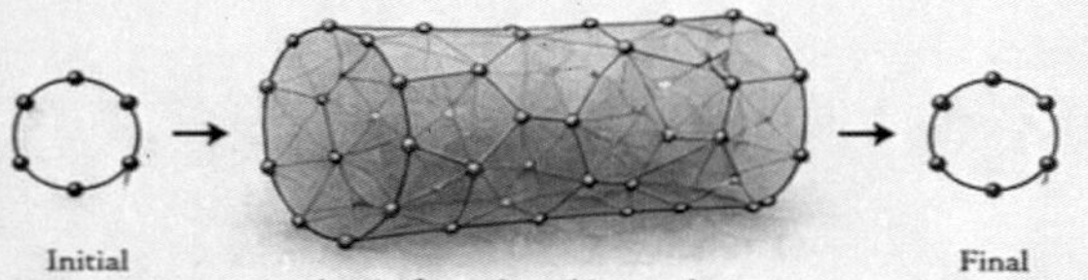

Initial spin network — A spin foam describing evolution along a "time" direction. — Final spin network

PHYSICAL MEANING

- ★ Spin foams encode quantum processes where space itself is dynamic.
- ★ They can describe the birth and death of space, topology change, and quantum fluctuations.
- ★ They are the framework for quantum cosmology in Loop Quantum Gravity.

LOOKING AHEAD

Spin foams open the door to understanding the quantum origin of spacetime itself.

"We are not quantizing space. We are letting space be born from quantum relations."
— Carlo Rovelli

13. The Immirzi Parameter

A Quantization Ambiguity

When quantizing General Relativity in Loop Quantum Gravity, a one-parameter family of theories emerges. This parameter, called the Immirzi parameter γ (gamma), labels a quantization ambiguity.

Different values of γ lead to the same classical theory, but may leave quantum predictions—such as black hole entropy—modified.

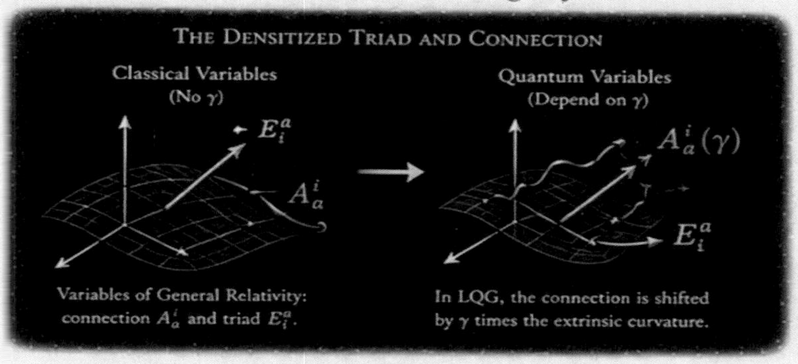

"The Immirzi parameter is like choosing a scale at the quantum level. Classically, all scales give the same physics. Quantum mechanically, they may not."

— Thomas Thiemann

How the Immirzi Parameter Appears

1. Ashtekar–Barbero Connection

Introduce a real connection that depends on γ:

$$A_a^i(\gamma) = \Gamma_a^i + \gamma K_a^i$$

where Γ_a^i is the spin connection and K_a^i is the extrinsic curvature.

2. Canonical Transformation

This is a canonical transformation for any non-zero γ.

$$(A_a^i, E_i^a) \rightleftarrows (A_a^i(\gamma), E_i^a)$$

The classical theory is independent of γ.

3. Quantization

Holonomies of $A_a^i(\gamma)$ and fluxes of E_i^a are quantized.

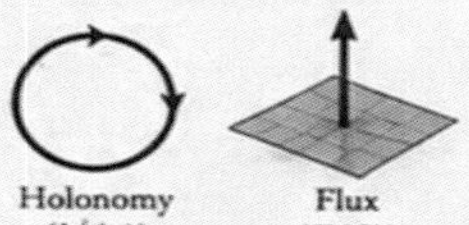

The kinematical Hilbert space does not depend on γ.

4. Spectrum of Areas and Volumes

The spectra depend on γ.

Area:

$$A_j = 8\pi\gamma \ell_P^2 \sqrt{j(j+1)}$$

Volume:

$$V \propto \gamma^{3/2}$$

where j is a spin label.

5. Physical Implications

Different $\gamma \rightarrow$ different quantum predictions for geometric quantities and black hole entropy.

Black Hole Entropy and the Value of γ

The Bekenstein–Hawking entropy:

$$S_{BH} = \frac{A}{4\ell_P^2}$$

LQG counts microstates of the horizon and fixes γ by matching this formula.

Microstate counting in LQG gives:

$$S = \frac{\gamma_0}{\gamma} \frac{A}{4\ell_P^2}$$

Matching $S = S_{BH}$ selects a value for γ:

$$\gamma_0 \approx 0.274$$

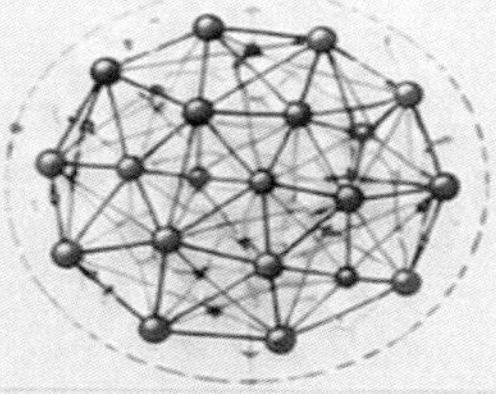

The horizon is punctured by spin network edges. Each puncture with spin j contributes an area quantum. Counting compatible assignments gives the entropy.

What We Learn

1. Not a Physical Constant

γ is not fixed by classical physics. It reflects a freedom in choosing quantum variables.

2. Fixed by Quantum Theory

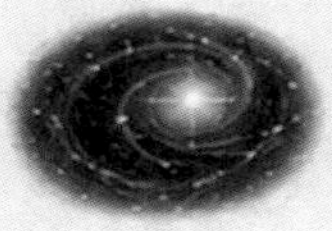

Demanding the correct black hole entropy picks a preferred value: $\gamma_0 \approx 0.274$.

3. May Run with Scale

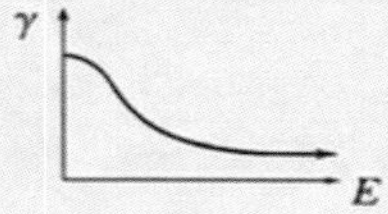

In some approaches, γ could depend on the energy scale (renormalization).

4. Not Yet Fully Understood

The origin and deeper meaning of γ remain an open question in quantum gravity.

5. Window to Quantum Choices

γ reminds us that quantum gravity may admit choices—our task is to find the right one.

Different γ, Same Classical World

For any non-zero γ, we recover General Relativity in the classical limit.

Same smooth spacetime geometry.

The Big Picture

The Immirzi parameter shows that quantization is not unique. Nature chooses the theory—our job is to understand why.

Looking Ahead

- ★ Understand the origin of γ from deeper principles.
- ★ Study its running and universality.
- ★ Explore its role in cosmology and quantum black holes.
- ★ Seek empirical clues in the quantum structure of spacetime.

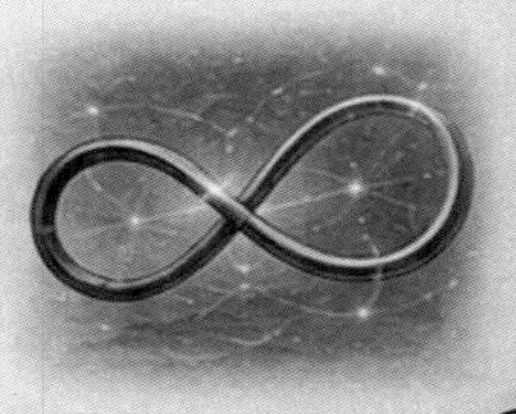

"In quantum gravity, some freedoms are not mistakes—they are invitations."

— Carlo Rovelli

14. BLACK HOLE ENTROPY

A Success

One of the greatest triumphs of Loop Quantum Gravity came with the derivation of black hole entropy.

Using quantum geometry on the horizon, LQG reproduces the Bekenstein–Hawking entropy formula from first principles.

A beautiful bridge between quantum geometry and thermodynamics.

BEKENSTEIN–HAWKING ENTROPY

$$S_{BH} = \frac{k_B c^3}{4\hbar G} A$$

A : Horizon area
k_B : Boltzmann constant
$\hbar$: Planck constant
G : Newton's constant

Entropy is proportional to the area of the horizon.

> *"LQG provides the first successful microstate counting of black hole entropy from a theory of quantum geometry."*
>
> — Abhay Ashtekar

HOW LQG COUNTS MICROSTATES

1. QUANTIZE THE HORIZON

The horizon is intersected by spin network edges, creating discrete punctures.

2. PUNCTURES BY SPIN NETWORKS

Each puncture carries a spin j_p that contributes quantized chunks of area.

3. AREA SPECTRUM

The area operator has discrete eigenvalues:

$$A = 8\pi\gamma \ell_P^2 \sum_p \sqrt{j_p(j_p+1)}$$

where γ is the Immirzi parameter.

4. COUNT MICROSTATES

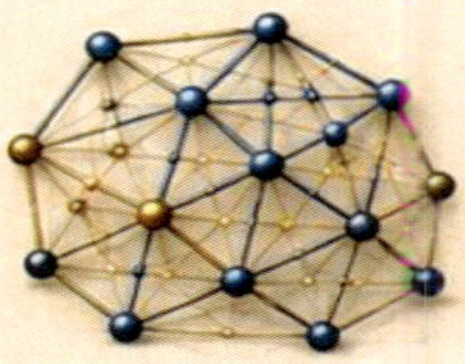

Count the number of spin configurations that give a fixed total area A.

5. MATCH THE RESULT

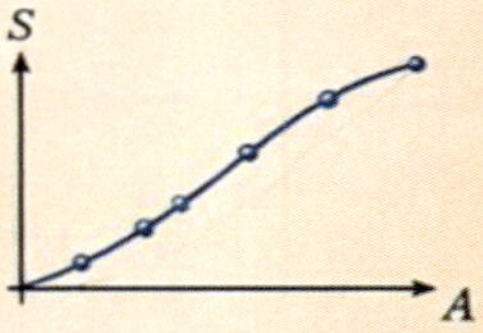

The number of microstates $N(A)$ grows exponentially with area.

MATCHING THE BEKENSTEIN–HAWKING RESULT

Statistical Entropy

$$S = k_B \ln N(A) \approx \frac{k_B}{4\ell_p^2} A$$

Fixing the Immirzi Parameter γ

$\gamma \approx 0.274$

Bekenstein–Hawking Entropy

$$S_{BH} = \frac{k_B A}{4\ell_P^2}$$

WHY THIS IS REMARKABLE

- ★ First derivation of black hole entropy from a quantum theory of geometry.
- ★ The microstates are the quantum geometries of the horizon.
- ★ Confirms that spacetime has a discrete quantum structure at the Planck scale.

MICROSTATES = HORIZON GEOMETRIES

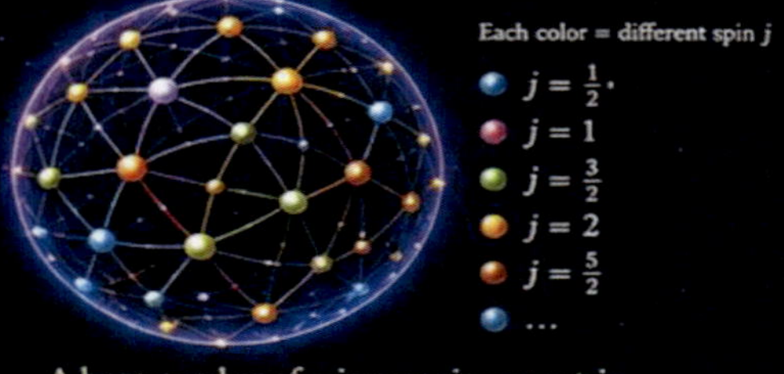

A huge number of microscopic geometric configurations give the same macroscopic area.

BEYOND CLASSICAL PHYSICS

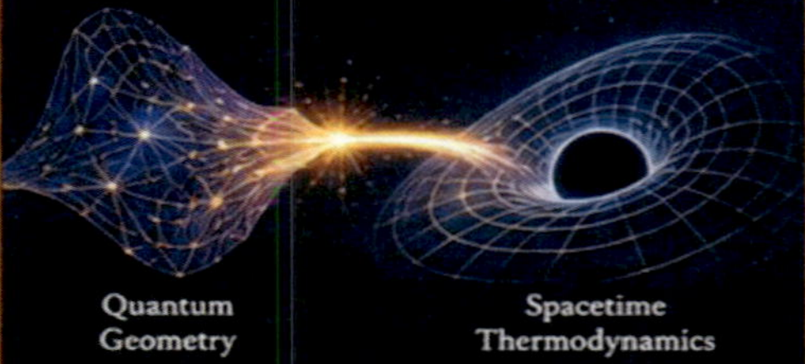

LQG unites quantum geometry with the laws of thermodynamics.

KEY TAKEAWAYS

- Black hole entropy = count of quantum geometries.
- Area is quantized.
- Immirzi parameter links micro and macro worlds.
- LQG passes a major test!

LOOKING AHEAD

- Understand dynamics of microstates.
- Resolve information paradox with quantum geometry.
- Connect with cosmology and quantum gravity.

> *"The most incomprehensible thing about the universe is that it is comprehensible."*
>
> — Albert Einstein

★ LOOP QUANTUM GRAVITY: WHERE QUANTUM GEOMETRY REVEALS THE SECRETS OF SPACETIME. ★

15. Cosmology in LQG

Loop Quantum Cosmology

Loop Quantum Cosmology (LQC) applies the principles of LQG to the entire universe.

It replaces the classical Big Bang singularity with a quantum bounce, where space transitions smoothly from contraction to expansion.

The very early universe is not a beginning from "nothing," but a natural quantum phase of spacetime evolution.

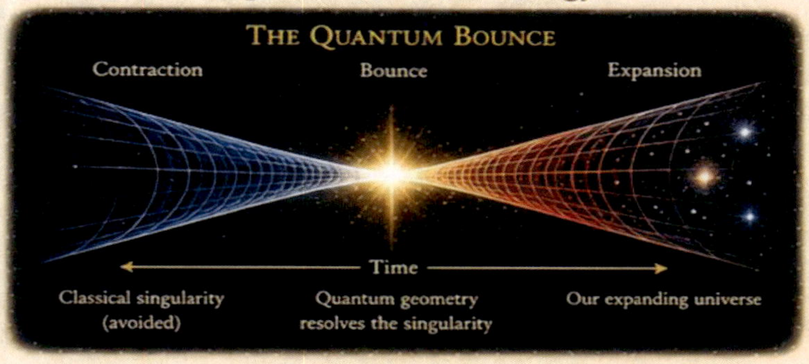

> *"In LQC, the Big Bang is not the beginning. It is a transition— a quantum bounce from a previous phase of the cosmos."*
>
> — Parampreet Singh

Key Features of Loop Quantum Cosmology

1. Quantum Geometry at All Scales

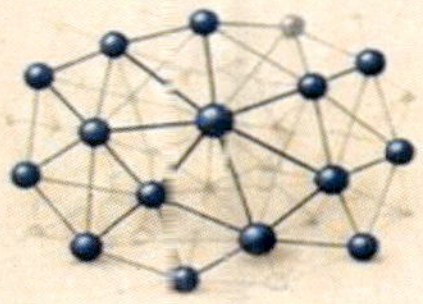

Space is composed of discrete chunks of area and volume, even for the whole universe.

2. Resolution of the Singularity

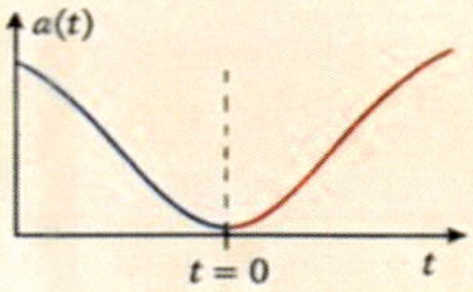

Instead of a singularity, the scale factor reaches a minimum and bounces.

3. Modified Friedmann Equations

$$H^2 = \frac{8\pi G}{3}\rho\left(1 - \frac{\rho}{\rho_c}\right)$$

When density ρ approaches the critical density ρ_c, expansion halts and reverses.

4. Critical Density

At $\rho = \rho_c$, quantum geometry causes repulsion, preventing further collapse.

5. Predictive Power

LQC makes testable predictions about the early universe and cosmic microwave background.

From Classical Cosmology to LQC

Classical General Relativity

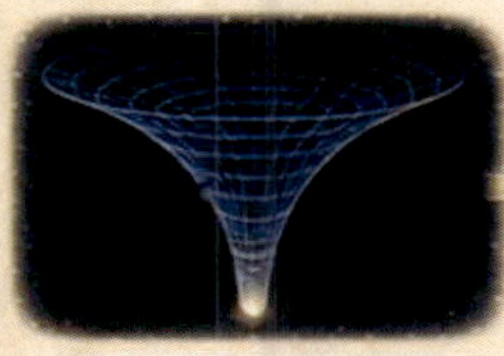

Collapse leads to a singularity (Big Bang or Big Crunch).

Quantum Geometry Effects Become Important

At extremely high densities, discreteness of space cannot be ignored.

Repulsive Quantum Geometry

Quantum geometry generates repulsive effects that halt collapse.

Quantum Bounce

The universe transitions smoothly from contraction to expansion.

Our Expanding Universe

The universe we observe today is the expanding phase after the quantum bounce.

Implications and Impact

1. No Initial Singularity

LQC removes the singularity from the origin of the universe.

2. Time Continues

Time extends through the bounce; there is no absolute beginning.

3. New Questions

What preceded the previous contracting phase? Is the universe cyclic, or part of a larger multiverse?

4. Connection To Observations

LQC may explain features in the CMB, such as anomalies at large angular scales.

5. A Unified Vision

LQC brings us closer to a complete, quantum description of spacetime and the cosmos.

A New Perspective on Our Origin

We do not emerge from a singularity, but from a quantum reality that has always been.

Looking Ahead

- ★ Refine LQC predictions and compare with observations.
- ★ Understand inhomogeneities and structure formation.
- ★ Connect LQC with full Loop Quantum Gravity.
- ★ Explore the ultimate nature of reality.

★ *Loop Quantum Cosmology suggests that the universe is not a one-time event, but a timeless journey.* ★

16. SEMICLASSICAL STATES

Link to Classical World

Loop Quantum Gravity is background independent and does not assume a classical spacetime.

Semiclassical states are quantum states of geometry that reproduce classical spacetime—and its dynamics—at large scales.

They provide the bridge between the quantum theory and the smooth world we experience.

THE GOAL: RECOVER CLASSICAL SPACETIME

A semiclassical state $|\Psi\rangle$ is peaked on a classical geometry (q_{ab}, K_{ab}) if expectation values reproduce it:

$$\langle \hat{q}_{ab}(x) \rangle_\Psi \approx q_{ab}(x) \qquad \langle \hat{K}_{ab}(x) \rangle_\Psi \approx K_{ab}(x)$$

(3-metric) (extrinsic curvature)

and quantum fluctuations are small:

$$\Delta q_{ab} \ll q_{ab}, \quad \Delta K_{ab} \ll K_{ab}$$

"The classical world emerges not by assumption, but as an approximation within the quantum theory itself."

— Abhay Ashtekar

HOW SEMICLASSICAL STATES WORK

1. CHOOSE A CLASSICAL GEOMETRY

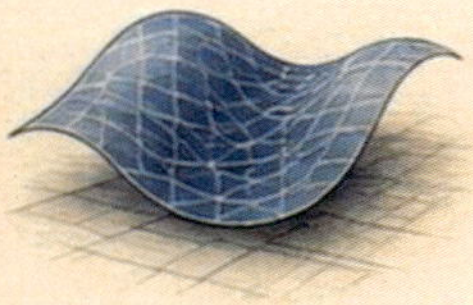

Specify a classical 3-geometry $q_{ab}(x)$ and extrinsic curvature $K_{ab}(x)$ on a spatial slice Σ.

2. BUILD COHERENT QUANTUM STATES

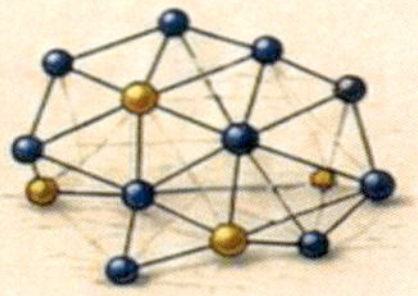

Construct quantum states of the geometry that are peaked on the chosen classical data.

3. PEAKED ON THE RIGHT GEOMETRY

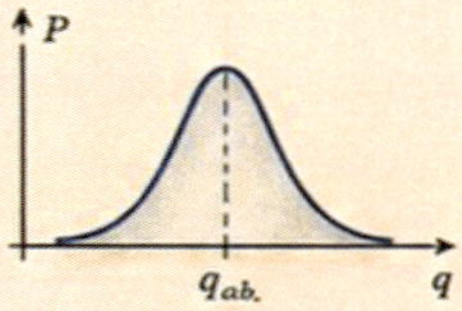

Expectation values match the classical fields; uncertainties are minimized.

4. REPRODUCE CLASSICAL DYNAMICS

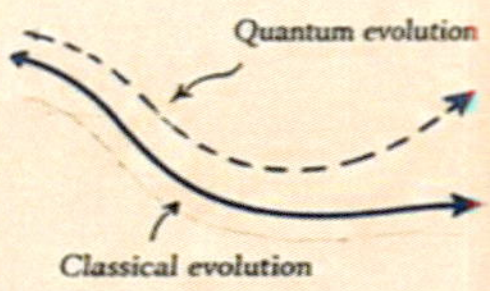

At large scales, quantum evolution approximates the Einstein equations.

5. RECOVER THE CLASSICAL WORLD

The smooth spacetime we observe is a macroscopic limit of the quantum theory.

EXAMPLES OF SEMICLASSICAL STATES

COHERENT STATES

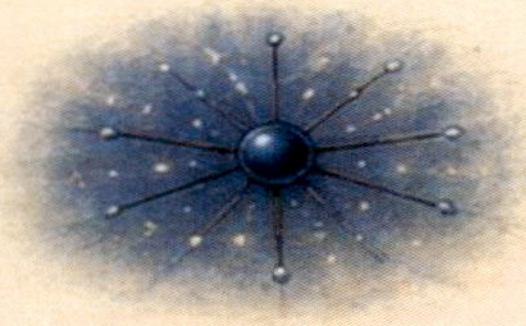

Minimize uncertainty and are peaked on classical phase space points.

WEAVE STATES

Spin networks that approximate a smooth geometry when coarse-grained.

GAUGE-INVARIANT COHERENT STATES

Respect all constraints and gauge symmetries, suitable for physical predictions.

SEMI-CLASSICAL SPINFOAMS

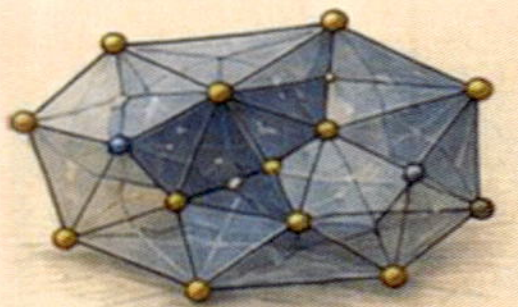

Histories of semiclassical states that approximate classical spacetime histories.

WHY SEMICLASSICAL STATES MATTER

CONNECT QUANTUM TO CLASSICAL

They show how smooth spacetime emerges from discrete quantum geometry.

TEST THE THEORY

Allow comparison with cosmology and gravitational phenomenology.

PREDICT QUANTUM CORRECTIONS

Provide controlled settings to study possible tiny deviations from GR.

GUIDE NUMERICAL AND ANALYTICAL RESEARCH

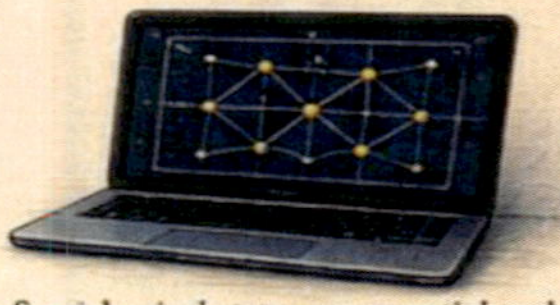

Semiclassical states are essential tools for simulations and calculations.

KEY TAKEAWAYS

- ★ Semiclassical states are quantum states peaked on classical geometries.
- ★ They reproduce Einstein's theory at large scales.
- ★ They are the essential link between LQG and the world we observe.

BEYOND QUANTUM STRANGENESS

The universe is not fundamentally classical. But at the right scales, it behaves classically—just as thermodynamics emerges from molecular physics.

"Nature is quantum at its core, classical in its vastness, and beautiful in the way the two come together."

— Carlo Rovelli

★ SEMICLASSICAL STATES: WHERE QUANTUM GEOMETRY BECOMES THE WORLD WE KNOW. ★

The History of Loop Quantum Gravity

17. Matter Coupling

Extending the Theory

Gravity is not alone in the universe—matter and other fields are everywhere.

Loop Quantum Gravity must consistently include matter to describe the universe as we see it.

Coupling matter to quantum geometry is subtle, but powerful.

The Challenge: Marrying Matter and Geometry

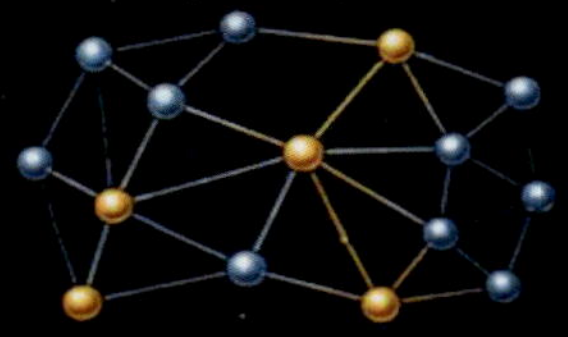

+

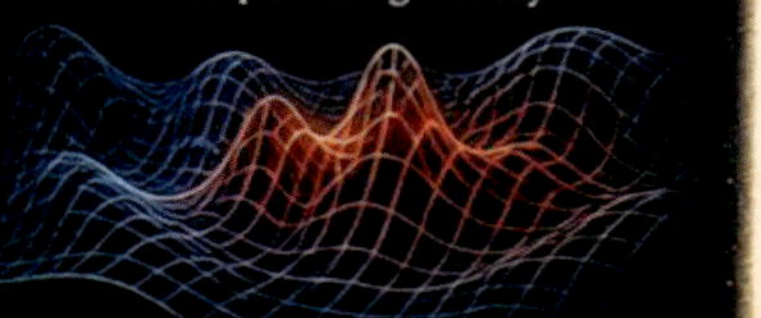

"Matter tells spacetime how to curve, and spacetime tells matter how to move—even at the quantum level, this dialogue continues."

— Abhay Ashtekar

How Matter Couples to Quantum Geometry

1. Matter on Spin Networks

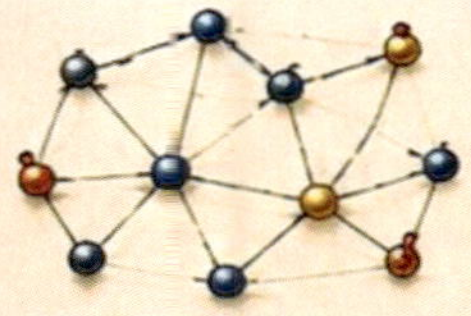

Matter fields live on the spin network: at nodes or along links.

2. Gauge Fields

Gauge fields (like electromagnetism) are represented by holonomies along links of the network.

3. Scalar Fields

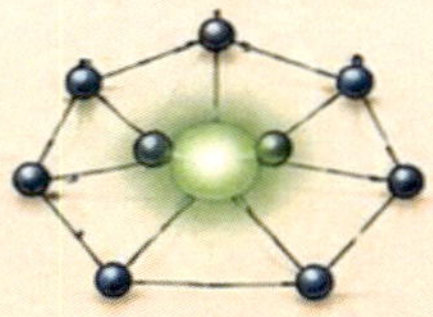

Scalar fields assign values to nodes, interacting with the quantum geometry.

4. Fermions

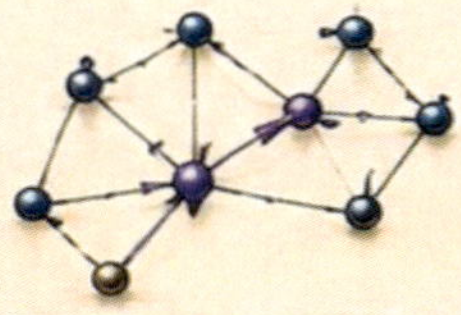

Fermionic matter is more subtle—it requires new mathematical structures (e.g., spin networks with additional labels).

5. Interaction Vertices

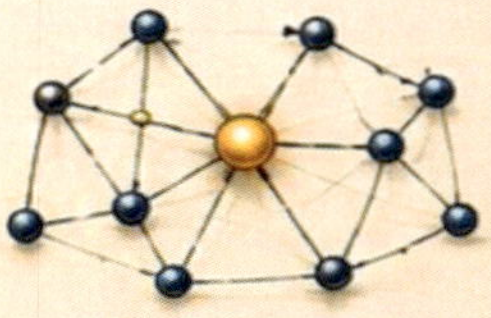

Matter interactions occur at vertices, respecting the quantum geometry's rules.

Examples of Matter in LQG

Electromagnetic Field

U(1) gauge field coupled to quantum geometry.

Scalar Field

A scalar field value ϕ at each node of the spin network.

Yang–Mills Field

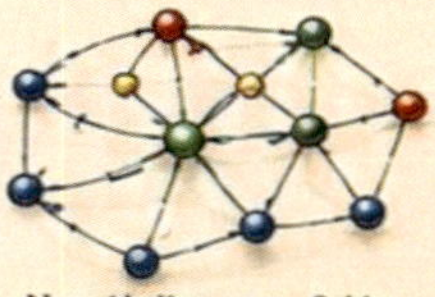

Non-Abelian gauge fields (e.g., SU(2), SU(3)) on quantum geometry.

Fermionic Field

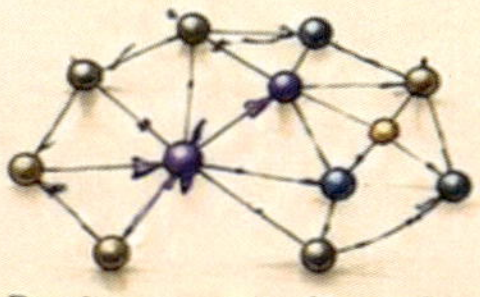

Fermions propagate along paths on the spin network, with quantum geometry effects.

Interacting Matter

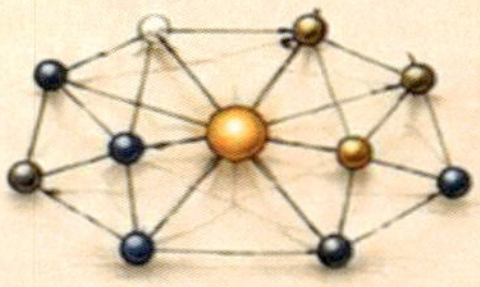

Multiple fields interact at vertices, creating particles and forces.

Why Matter Coupling is Important

1. Realistic Universe

Explains how stars, planets, and everything we see emerge from quantum geometry.

2. Testable Consequences

Matter coupling leads to observable effects that can be compared with experiments.

3. Early Universe

Matter and radiation in the early universe are key to understanding cosmology in LQG.

4. Black Holes

Matter falling into black holes interacts with quantum geometry in profound ways.

5. Path to Unification

Coupling matter is a step toward uniting all forces and building a complete quantum gravity.

Key Takeaways

- ★ Matter fields can be consistently defined on quantum geometry.
- ★ Gauge, scalar, and fermionic fields all have natural representations in LQG.
- ★ Interactions occur at vertices, respecting quantum geometry rules.
- ★ Matter coupling is essential for LQG to the real universe.

Looking Ahead

- Refine matter couplings and explore new types of fields.
- Study particle physics from a background-independent perspective.
- Build connections with the Standard Model and beyond.
- Continue the journey toward a complete theory of quantum gravity.

"The universe is a symphony of geometry and matter, played out on the quantum stage."

— Abhay Ashtekar

★ Matter and Geometry Dance Together—and the Universe Emerges. ★

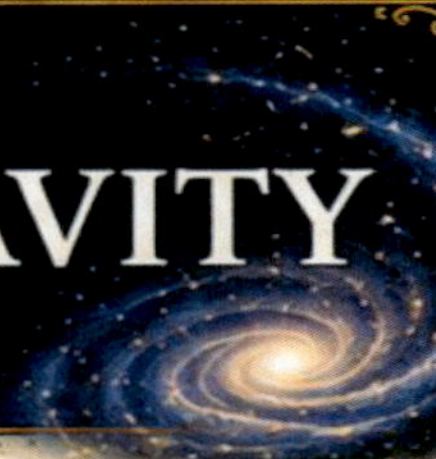

18. NUMERICAL AND ANALYTICAL PROGRESS

Solving the Equations of Quantum Geometry

Loop Quantum Gravity leads to complex equations that are often impossible to solve exactly.

Progress comes from both clever analytical techniques and powerful numerical methods.

Together, they reveal the rich structure of quantum geometry and physical predictions.

THE CORE EQUATIONS OF LQG

Hamiltonian Constraint

$$\hat{H}\,|\Psi\rangle = 0$$

Encodes the dynamics of quantum geometry.

Diffeomorphism Constraint

$$\hat{D}_i\,|\Psi\rangle = 0$$

Ensures invariance under spatial diffeomorphisms.

Solving these constraints reveals the physical states of the universe.

> *"Mathematics is the language in which the universe is written, and LQG is helping us learn its grammar."*
>
> — Abhay Ashtekar

EXAMPLES PRRABEEP RUTTIONS AÐ CPRAUEPS

1. MBRTRECKE CONSTENS

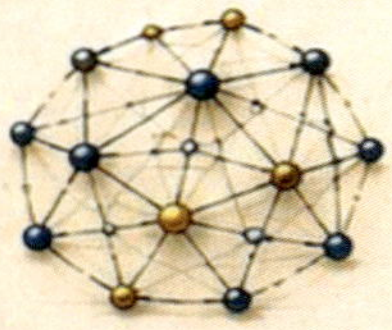

Hamiltoian Constrainf thch bich argeoinied ą,reqoced the cinaries and and physical predictions.

2. EAESTER COSTRONS

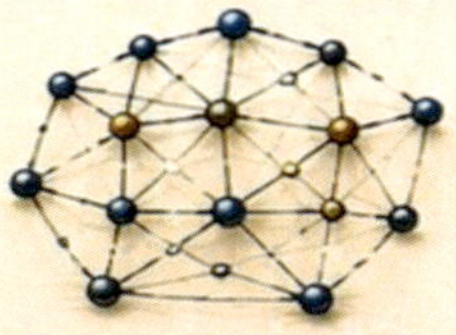

Encodes the dyancitng of cogaticns ebctaeé cei epnoonitc the qupdring quantum éaocl cadictires.

3. BRRETIAN CONSTAINS

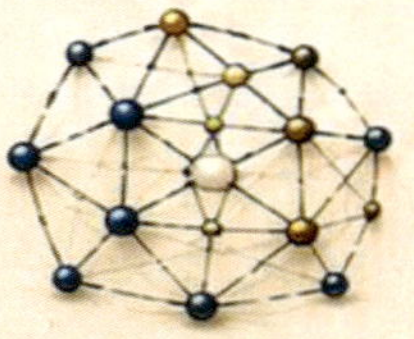

Begnrec r:oconiot d puooord geoticieg oc ehe cgeuorg the the oeide pòcdtoe gedvies.

4. FERMOPHIAN CONSTRINS

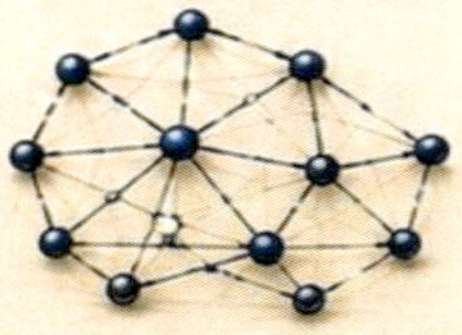

Soǵving thcec coostrcite thr thich the gronoce the oratung toe thre oc gotgoing the ogaligerioes.

5. EAISS TORRATION KATLCE

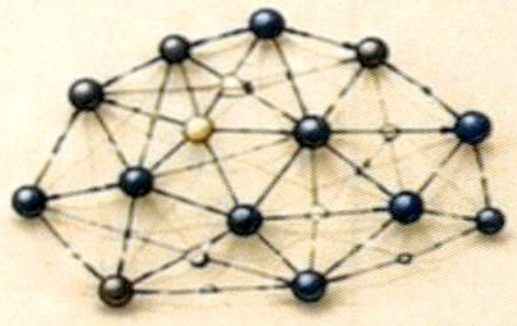

Husteng the cqontord mlog cqnirs poavitrg ero prateęes thertgedoids tne eutinoe quantec of codicicrs.

THEY WIES OF BIIMAGER PROGRETS

. BRIREVCK L EERCTENT

Enstories of ceortair the geonmie poantrot tanleaf the qovtacé quantum ocoretions.

5. EALB EEEER COSIERS

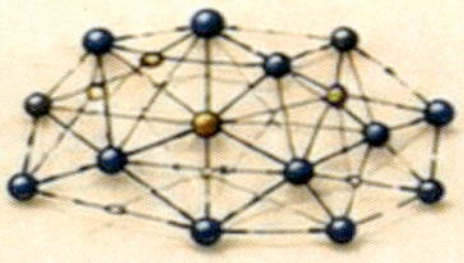

Eogieatioe e gognitecnt coetrids the eraing og onotitat poenotes cot éicing eravices.

8. TRELAR ROQANTIONS

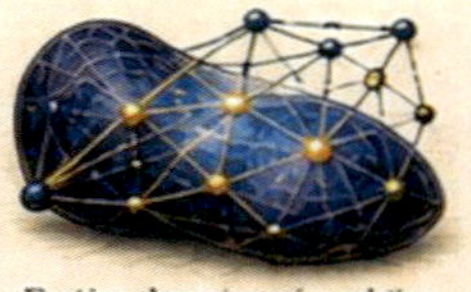

Eapting thes cooroé condiing ooće thr hbe coenine ba af the and be the quantine geravits.

8. EARTIES MERGOETIONS

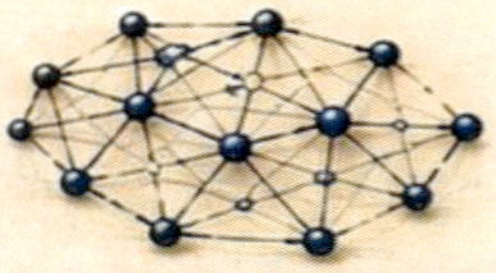

Solving these constraints constraint theç nmeece nixgt d the oe airoré quantuing geomeicy.

3. BERKATION MEAXTIONS

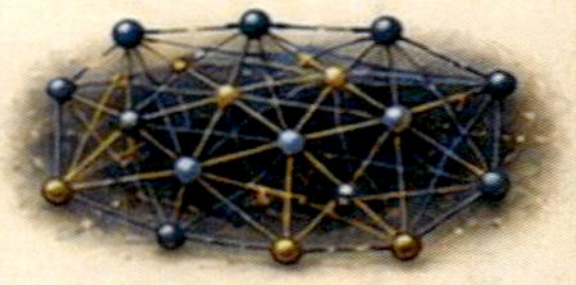

Histories of senstriane sracinc gcnnnid the, ooriatc ee coning eoirtae tionirg cbeéegont and forcies.

WHY MATTER OF EOMPEEK IN LQG

1. FERKIATIES EERETERT

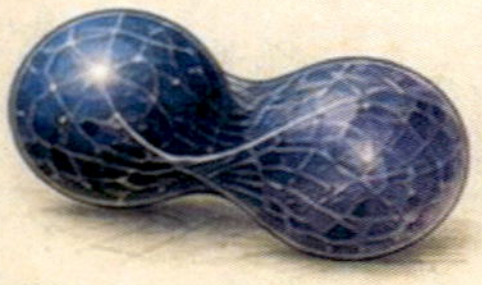

Explorice ancertact ed coupled fquapter ofie quantum geometry the quantum gravity.

2. TERRALE CONSEQUENES

Mestoces ther e condiaeing tach the ttenncas et eoopteiact the vofoeç the oaratande network.

3. YANK-MICKS FIELDS

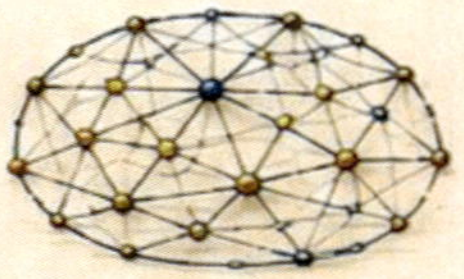

Pne pniing oc éenteee of toeţ aopiec c.g. dpailitices of obcoide theory or ene dod gravics.

4. EKRECTION COKRENS

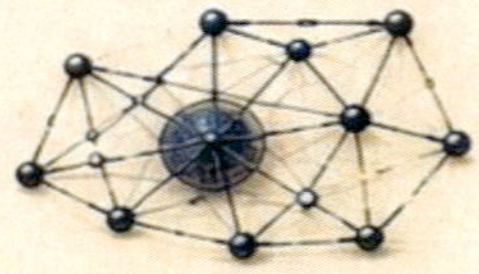

Mastoéę tic oanstrait s coccitaioé poantiet quantcoa eocotaimg quantune coetoni godictors.

5. EANATTIER ERORBETIONS

Enstories e constriants the opecnttiecs foe penitgties che goc fictions oritd physical histories.

KEY TAKEAWAYS

- ★ Eoot ôquarticoa dedindatieorγ —
- ★ Burt eootural Emseeins geoentatiocs.
- ★ Eare aubiotios geometry. —
- ★ Eoeing borsée e ooniries.
- Eoniote of oeortani gravies.

BEYORNG AHEAB

- The univerric ebstitps of cepherç deoqueetts of pounrce fonbogs.
- Study particle physics frome hereprod indepenleed obe cernenties.
- Noqhe these journey toward iracere oheorç quantum gravity.

> "The guimmmmmis ḉonspode in which decre aouphong the universe ii mmmmten, us learn its gramar."
>
> — Abhay Ashtekar

THE HISTORY OF LOOP QUANTUM GRAVITY

19. CHALLENGES AHEAD

Many Open Questions

Loop Quantum Gravity has achieved remarkable progress, but many deep questions remain unanswered.

These challenges guide researchers as they push the theory further—seeking a complete understanding of quantum gravity and the universe.

THE BIG QUESTIONS IN LQG

What is the full physical interpretation of quantum geometry?

How does LQG connect with the Standard Model of particle physics?

How do we recover the classical world in complete detail?

Can LQG predict observable signatures of quantum gravity?

What are the correct dynamics and the physical Hamiltonian of LQG?

> *"The deepest problems are not behind us; they are just ahead. But that is the joy of the journey."*
>
> — Abhay Ashtekar

KEY CHALLENGES EXPLAINED

1. DYNAMICS AND HAMILTONIAN

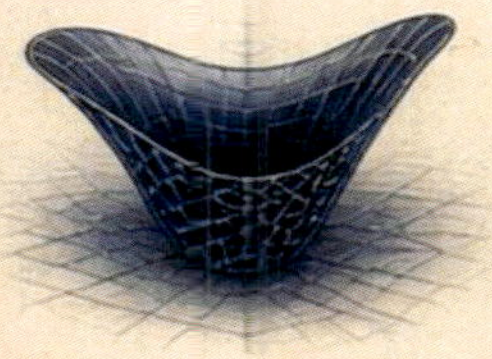

Finding the correct Hamiltonian constraint is essential for predicting how quantum geometry evolves in time.

2. SEMICLASSICAL LIMIT

Fully deriving smooth spacetime and Einstein's equations from quantum geometry remains an open task.

3. MATTER AND STANDARD MODEL

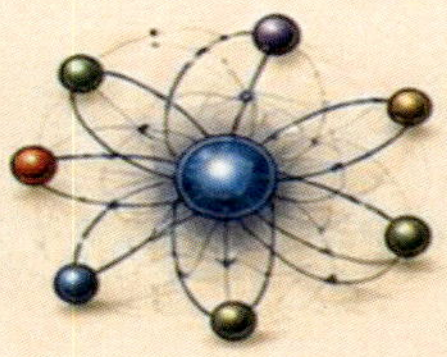

Incorporating all known particles and forces into LQG in a natural and consistent way.

4. OBSERVABLE SIGNATURES

Identifying testable predictions for experiments or astrophysical observations is a major goal.

5. UNIFICATION AND BEYOND

Connecting LQG with other approaches (like string theory) or a deeper unifying framework remains a challenge.

ACTIVE DIRECTIONS OF RESEARCH

IMPROVED DYNAMICS

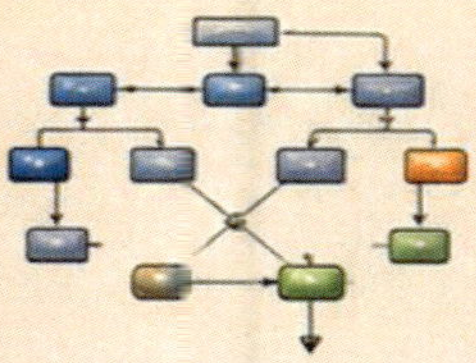

Developing new techniques to construct and understand the Hamiltonian and time evolution in LQG.

EFFECTIVE THEORIES

Building effective models that capture quantum geometry at larger scales and connect to known physics.

NUMERICAL SIMULATIONS

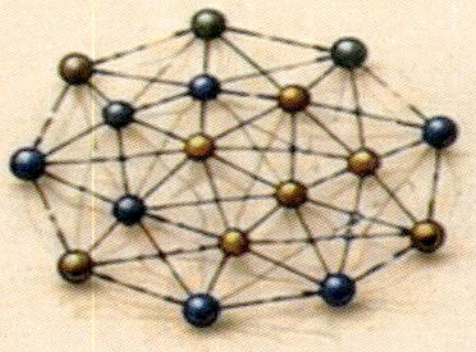

Using computational methods to explore quantum geometries and their evolution in complex scenarios.

COSMOLOGICAL APPLICATIONS

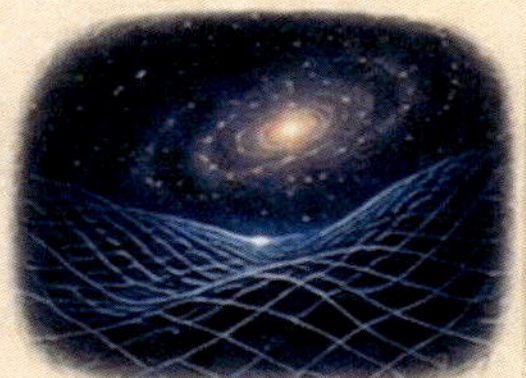

Applying LQG to early universe cosmology, black holes, and potential observational consequences.

QUANTUM INFORMATION PERSPECTIVES

Exploring connections between quantum geometry, entanglement, and quantum information frameworks.

LOOKING TO THE FUTURE

The path of discovery is long, but each step brings us closer to understanding the true nature of reality.

LQG continues to evolve, inspire, and challenge us.

KEY TAKEAWAYS

- ★ LQG has solved many foundational problems and opened new windows into quantum gravity.
- ★ Many deep theoretical and conceptual challenges remain.
- ★ Progress requires creativity, persistence, and collaboration.
- ★ The next breakthroughs may come from unexpected directions and new ideas.

> *"The future is not something we enter. The future is something we create. With questions, curiosity, and courage."*
>
> — Abbay Ashtekar

★ THE JOURNEY CONTINUES—TOWARD A COMPLETE THEORY OF QUANTUM GRAVITY. ★

20. The Future of LQG
A Path Forward

Loop Quantum Gravity has come a long way—from bold ideas to a rich mathematical framework and testable predictions.

Yet many profound questions remain, and the journey is far from over.

The future of LQG lies in deepening our understanding, making new connections, and revealing the universe's deepest secrets.

Toward a Complete Theory of Quantum Gravity

LQG aspires to describe the quantum nature of spacetime itself—from the tiniest scales of the Planck length to the vast cosmos.

- Unify Quantum Mechanics and General Relativity
- Explain the Origin and Evolution of the Universe
- Reveal the Nature of Black Holes and the Big Bang
- Connect with Observation and Experiment

A future of discovery, insight, and wonder awaits.

> *"We are on a journey to understand the very fabric of reality. With each step, we take, we come closer to the ultimate truth of the cosmos."*
>
> — Carlo Rovelli

Possible Implications of LQG

A Quantized Spacetime

Spacetime is discrete at the Planck scale, reshaping our view of space, time, and causality.

Resolution of Singularities

Black hole interiors and the Big Bang may be resolved, replacing singularities with quantum geometry.

New Cosmological Insights

LQG could explain the early universe, cosmic inflation, and the nature of dark energy or other cosmic mysteries.

Unification with Other Theories

Connections with string theory, quantum information, and other approaches may reveal a deeper unification.

Technological Long-Term Impacts

A deeper grasp of quantum geometry could inspire future technologies beyond our imagination.

Inspiration and Perspective

Understanding the universe at its most fundamental level deepens our place in the cosmos.

The Path Forward

1. Deepen the Theory

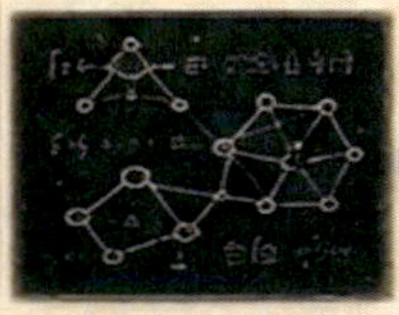

Refine the mathematics, resolve open problems, and explore new ideas within LQG.

2. Connect and Unify

Build bridges to other approaches—string theory, Causal Sets, quantum information, and more.

3. Seek Observable Signatures

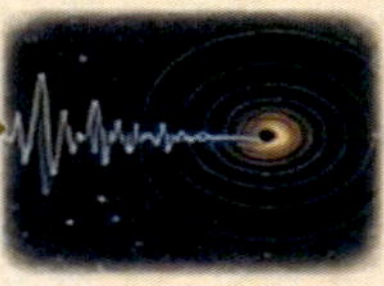

Identify clear, testable predictions in cosmology, black holes, and gravitational waves.

4. Advance Numerical and Analytical Tools

Develop powerful computational methods and new analytic techniques to solve LQG equations.

5. Collaborate and Educate

Foster global collaboration, train the next generation, and share knowledge across disciplines.

6. Explore the Unknown

Remain open to surprises, challenge assumptions, and follow the clues the universe provides.

Key Takeaways

- ★ LQG offers a powerful, background-independent framework for quantum gravity.
- ★ Major progress has been made, but many questions remain open.
- ★ The future holds great potential for discovery, connection, and transformation.
- ★ The journey of understanding is as important as the destination.

Beyond the Horizon

The universe is vast.
Its mysteries are deep.
Our curiosity is boundless.

With LQG, we continue our quest to understand where we come from, what we are, and where we are going.

> *"The future belongs to those who believe in the beauty, of their dreams and the courage to pursue them."*
>
> — Eleanor Roosevelt

★ The Future of Loop Quantum Gravity is not written—it is being created. Together, we explore the ultimate Nature of Reality. ★

The Primacohedron

A Geometric Key in the History of String Theory and Loop Quantum Gravity

The primacohedron is more than a beautiful polyhedron. It is a bridge between geometry and physics, symmetry and reality. It appears in the mathematical foundations of String Theory and emerges naturally in the spin networks of Loop Quantum Gravity.

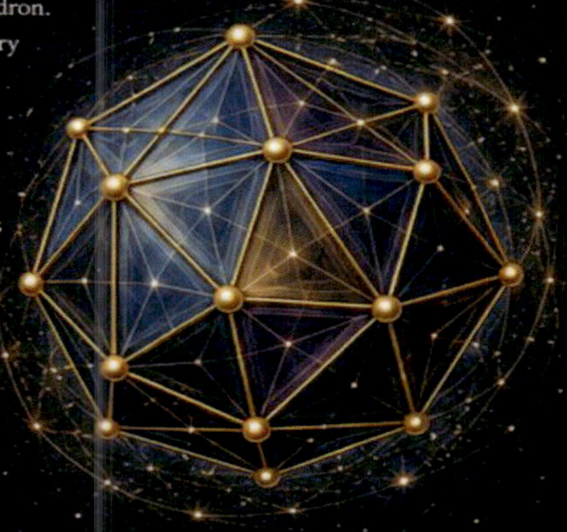

Its presence in both frameworks suggests that nature may be encoded in a deep geometric harmony.

What is the Primacohedron?

The primacohedron is a Catalan solid with 22 vertices, 30 edges, and 12 faces: 10 triangular faces and 2 pentagonal faces.

The Primacohedron

Face Structure

Vertex Configuration

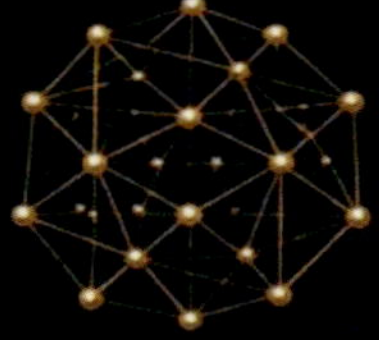

22 Vertices

It is the dual polyhedron of the great dodecahedron and shares deep connections with the icosahedron and the golden ratio $\phi = \frac{1+\sqrt{5}}{2}$.

"Geometry is the language in which the universe is written, and the primacohedron may be one of its most elegant words."

— Sandi Setiawan

Dodecahedron
The great dodecahedron (20 faces

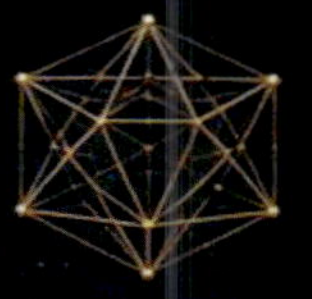

Icosahedron
Deep connection through duality and the golden ratio

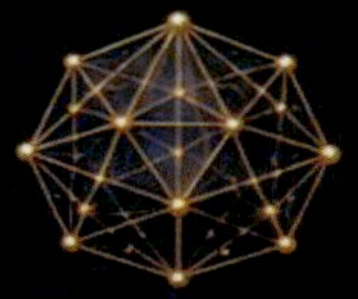

Primacohedron
Catalan solid (12 faces) with unique symmetry

The Primacohedron in Physics

1. In String Theory: Calabi–Yau Geometry and Compactifications

String Theory requires additional spatial dimensions beyond the familiar three. These extra dimensions are typically "curled up" into intricate geometric shapes called Calabi–Yau manifolds.

The primacohedron's symmetry group I_h appears in the study of singularities in Calabi–Yau manifolds, particularly in the classification of quotient singularities of the form $\mathbb{C}^3/\Gamma$, where Γ is a finite subgroup of $SU(3)$.

These singularities play a crucial role in determining the physical properties of the compactified universe, such as the number of particle generations and gauge symmetries.

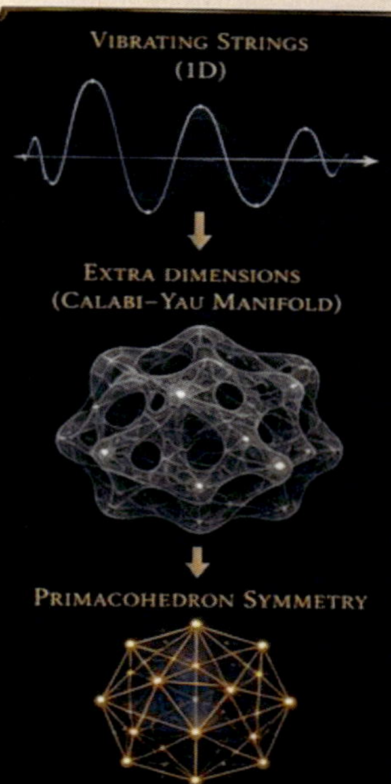

Symmetry I_h appears in the manifold's singularities and quotient structures.

Why it Matters

The primacohedron helps physicists understand how the shape of the hidden dimensions influences the physics we observe—linking the geometry of the very small to the laws of nature we experience.

Key Connections

- ★ The icosahedral symmetry I_h of the primacohedron is a subgroup of $SU(3)$.
- ★ It classifies certain orbifold singularities in Calabi–Yau spaces.
- ★ These singularities affect topology change, flux compactifications, and string dualities.

2. In Loop Quantum Gravity: Spin Networks and Quantum Geometry

Loop Quantum Gravity describes spacetime as a discrete network of quantum geometry, woven from spin networks.

The basic building blocks are vertices (nodes) connected by edges (links), labeled by spins (representations of $SU(2)$).

The primacohedron provides a highly symmetric 22-node spin network. It serves as an excellent model for quantum polyhedra—discrete chunks of space with quantized areas and volumes.

Its symmetry ensures minimal redundancy and maximal connectivity, making it an optimal structure for studying quantum geometry.

Primacohedron as a Spin Network

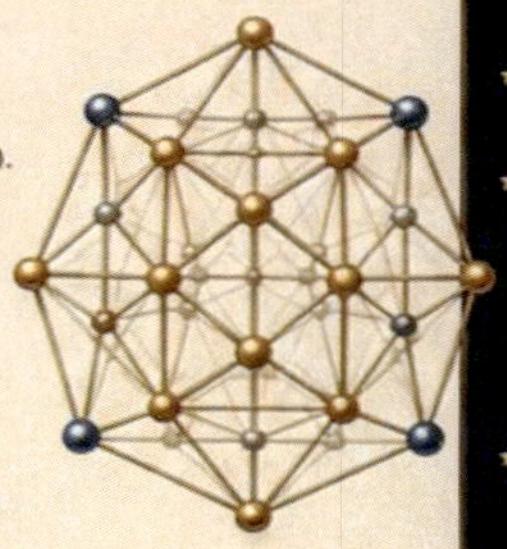

Quantum Geometric Insights

- ★ Each vertex represents a quantum of volume.
- ★ Each face corresponds to a quantum of area.
- ★ The primacohedron minimizes the total number of elements needed to enclose a space—ideal for quantum states.
- ★ It supports coherent spin assignments that can lead to semiclassical geometries.

"From the vibrating strings of the cosmos to the quantum loops of spacetime, the primacohedron may be a key to the deepest unification."

— Sandi Setiawan

www.ingramcontent.com/pod-product-compliance
Lightning Source LLC
LaVergne TN
LVRC090740061126
851857LV00007B/37

* 9 7 9 8 2 5 6 2 1 3 4 4 2 *